Holt Geometry

Chapter 4 Resource Book

HOLT, RINEHART AND WINSTON

A Harcourt Education Company

Orlando • Austin • New York • San Diego • London

Printed in the United States of America

ISBN 0-03-042777-0

2 3 4 5 6 7 8 9 862 09 08 07 06

Contents

Blackline Masters

Parent Letter 1
4-1 Practice A, B, C 3
4-1 Reteach 6
4-1 Challenge 8
4-1 Problem Solving 9
4-1 Reading Strategies 10
4-2 Practice A, B, C 11
4-2 Reteach 14
4-2 Challenge 16
4-2 Problem Solving 17
4-2 Reading Strategies 18
4-3 Practice A, B, C 19
4-3 Reteach 22
4-3 Challenge 24
4-3 Problem Solving 25
4-3 Reading Strategies 26
4-4 Practice A, B, C 27
4-4 Reteach 30
4-4 Challenge 32
4-4 Problem Solving 33
4-4 Reading Strategies 34
4-5 Practice A, B, C 35
4-5 Reteach 38
4-5 Challenge 40
4-5 Problem Solving 41
4-5 Reading Strategies 42
4-6 Practice A, B, C 43
4-6 Reteach 46
4-6 Challenge 48
4-6 Problem Solving 49
4-6 Reading Strategies 50
4-7 Practice A, B, C 51
4-7 Reteach 54
4-7 Challenge 56
4-7 Problem Solving 57
4-7 Reading Strategies 58
4-8 Practice A, B, C 59
4-8 Reteach 62
4-8 Challenge 64
4-8 Problem Solving 65
4-8 Reading Strategies 66
Answers 67

Date______________

Dear Family,

In this chapter, your child will learn about the different types of triangles and the ways to determine triangle congruence.

Your child will learn to classify the different types of triangles, based on side lengths and angle measurements. This table outlines the different types of triangles your child will be responsible for knowing:

Type of Triangle	Example
acute	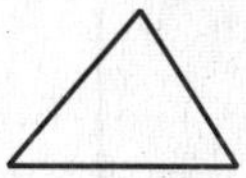
equiangular	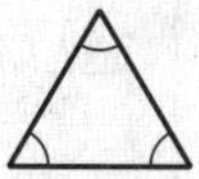
right	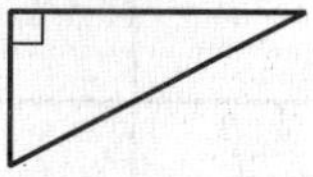
obtuse	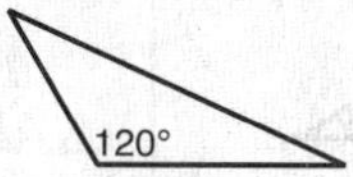 120°
equilateral	
isosceles	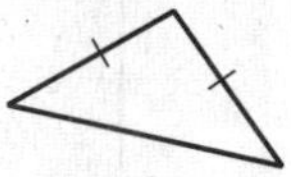
scalene	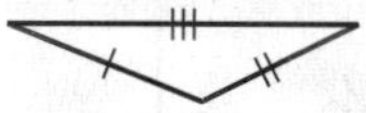

Your child will develop an understanding of the angle relationships in triangles. For example:

In a right triangle, the angles must add up to 180°. This means that if $m\angle Z = 90°$, then $m\angle X + m\angle Y = 90°$. Therefore $\angle X$ and $\angle Y$ are complementary angles.

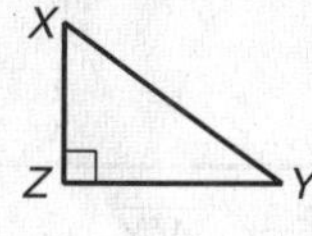

In an equiangular triangle, each angle must be equal to 60°.

Another example of angle relationships that your child will discover is the fact that the measure of an exterior angle of a triangle is equal to the sum of the measures of its remote interior angles.

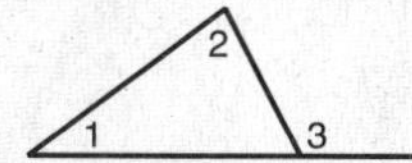

This means that $m\angle 3 = m\angle 1 + m\angle 2$.

Your child will use a series of theorems and postulates to describe triangle congruence. These are summarized below:

Postulate/Theorem	Example	Conclusion
Side-Side-Side Congruence If three sides of one triangle are congruent to three sides of another triangle, then the triangles are congruent.	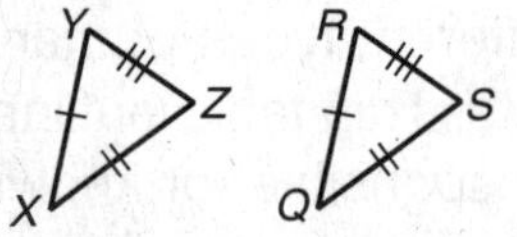	$\triangle XYZ \cong \triangle QRS$
Side-Angle-Side Congruence If two sides and the included angle of one triangle are congruent to two sides and the included angle of another triangle, then the triangles are congruent.	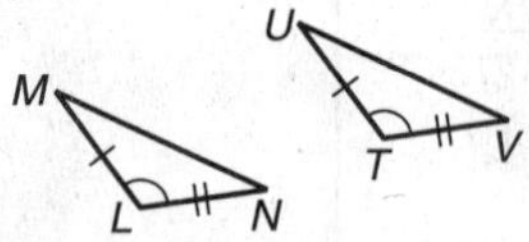	$\triangle LMN \cong \triangle TUV$
Angle-Side-Angle Congruence If two angles and the included side of one triangle are congruent to two angles and the included side of another triangle, then the triangles are congruent.	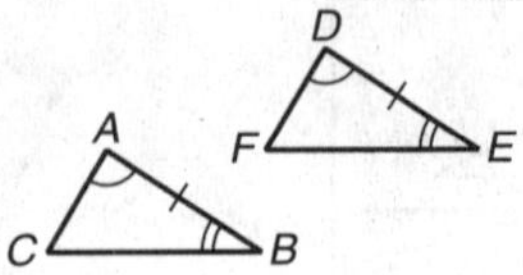	$\triangle ABC \cong \triangle DEF$
Angle-Angle-Side Congruence If two angles and a nonincluded side of one triangle are congruent to the corresponding angles and nonincluded side of another triangle, then the triangles are congruent.	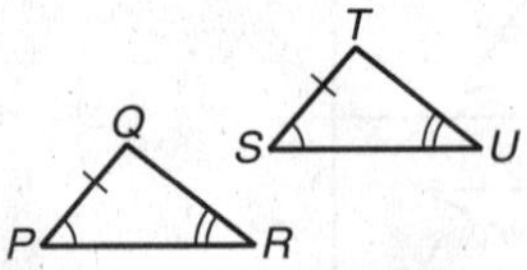	$\triangle PQR \cong \triangle STU$
Hypotenuse-Leg Congruence If the hypotenuse and a leg of a right triangle are congruent to the hypotenuse and a leg of another right triangle, then the triangles are congruent.	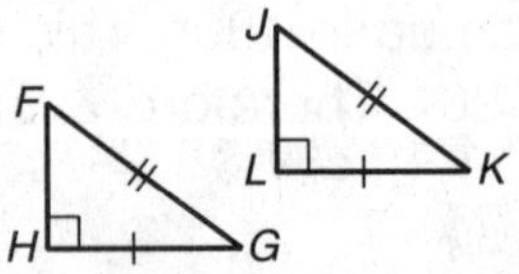	$\triangle FGH \cong \triangle JKL$
Corresponding Parts of Congruent Triangles are Congruent. (CPCTC)	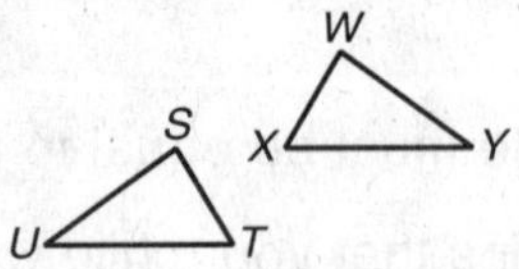	$\overline{ST} \cong \overline{WX}$ $\angle S \cong \angle W$ $\overline{SU} \cong \overline{WY}$ $\angle T \cong \angle X$ $\overline{TU} \cong \overline{XY}$ $\angle U \cong \angle Y$

For additional resources, visit *go.hrw.com* and enter the keyword MG7 Parent.

Name ______________________ Date ____________ Class ____________

LESSON 4-1

Practice A
Classifying Triangles

Match the letter of the figure to the correct vocabulary word in Exercises 1–4.

1. right triangle ________
2. obtuse triangle ________
3. acute triangle ________
4. equiangular triangle ________

A. 
B. 
C. 
D.  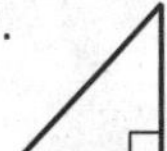

Classify each triangle by its angle measures as acute, equiangular, right, or obtuse. (*Note:* Give two classifications for Exercise 7.)

5. 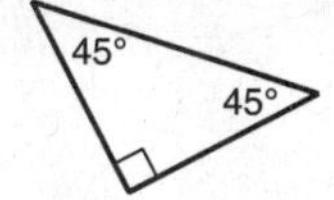

____

6. 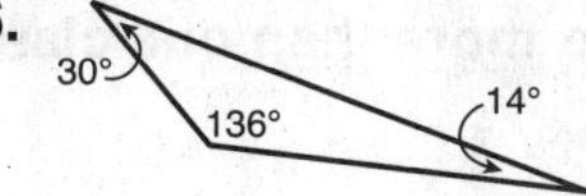

____

7. 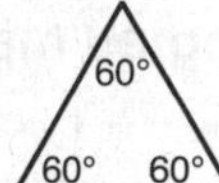

____

For Exercises 8–10, fill in the blanks to complete each definition.

8. An isosceles triangle has ________________ congruent sides.
9. An ________________ triangle has three congruent sides.
10. A ________________ triangle has no congruent sides.

Classify each triangle by its side lengths as equilateral, isosceles, or scalene. (*Note:* Give two classifications in Exercise 13.)

11. 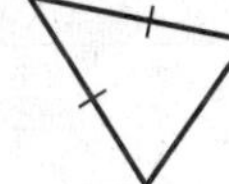

____

12. 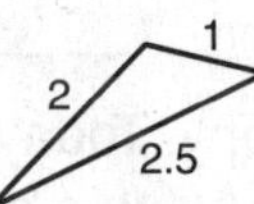

____

13. 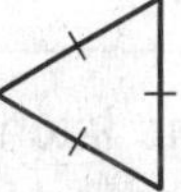

____

Find the side lengths of the triangle.

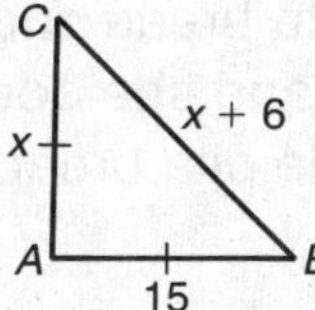

14. $AB =$ ____________ $AC =$ ____________ $BC =$ ____________

15. The New York City subway is known for its crowded cars. If all the seats in a car are taken, passengers must stand and steady themselves with railings or handholds. The last subway cars designed with steel hand straps were the "Redbirds" made in the late 1950s and early 1960s. The figure gives the dimensions of one of these triangular hand straps. How many hand straps could have been made from 99 inches of steel?

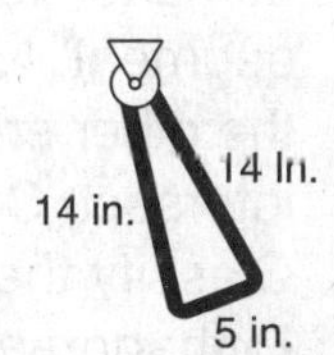

____

Name ______________________ Date ____________ Class ____________

Practice B
Classifying Triangles

Classify each triangle by its angle measures. (*Note:* Some triangles may belong to more than one class.)

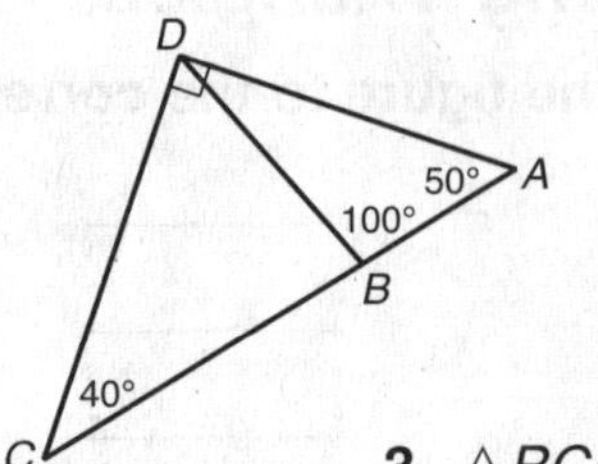

1. $\triangle ABD$ 2. $\triangle ADC$ 3. $\triangle BCD$

______________________ ______________________ ______________________

Classify each triangle by its side lengths. (*Note:* Some triangles may belong to more than one class.)

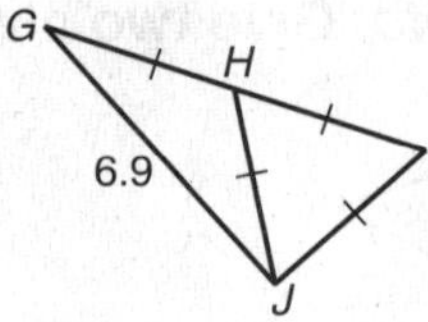

4. $\triangle GIJ$ 5. $\triangle HIJ$ 6. $\triangle GHJ$

______________________ ______________________ ______________________

Find the side lengths of each triangle.

7.

R, P, Q; $3x - 0.4$; $x + 0.1$; $x + 1.4$

8. 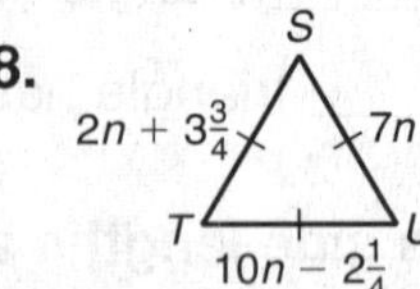

______________________ ______________________

9. Min works in the kitchen of a catering company. Today her job is to cut whole pita bread into small triangles. Min uses a cutting machine, so every pita triangle comes out the same. The figure shows an example. Min has been told to cut 3 pita triangles for every guest. There will be 250 guests. If the pita bread she uses comes in squares with 20-centimeter sides and she doesn't waste any bread, how many squares of whole pita bread will Min have to cut up?

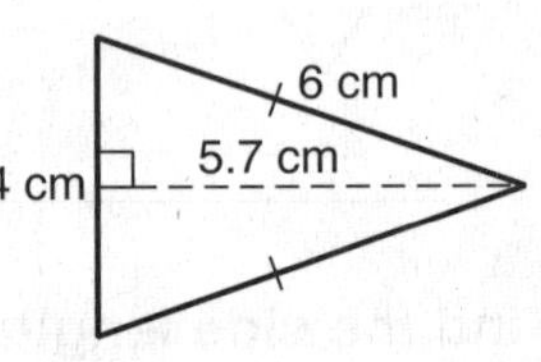

____

10. Follow these instructions and use a protractor to draw a triangle with sides of 3 cm, 4 cm, and 5 cm. First draw a 5-cm segment. Set your compass to 3 cm and make an arc from one end of the 5-cm segment. Now set your compass to 4 cm and make an arc from the other end of the 5-cm segment. Mark the point where the arcs intersect. Connect this point to the ends of the 5-cm segment. Classify the triangle by sides and by angles. Use the Pythagorean Theorem to check your answer.

____

Name ______________________ Date __________ Class __________

LESSON 4-1

Practice C
Classifying Triangles

The figure shows the side of a railway bridge. Steel girders make up the triangular pattern along the side and top. Each triangle is an equilateral triangle. In an equilateral triangle, the segment that shows the height also bisects the side it is perpendicular to. The height of the bridge is 18 feet.

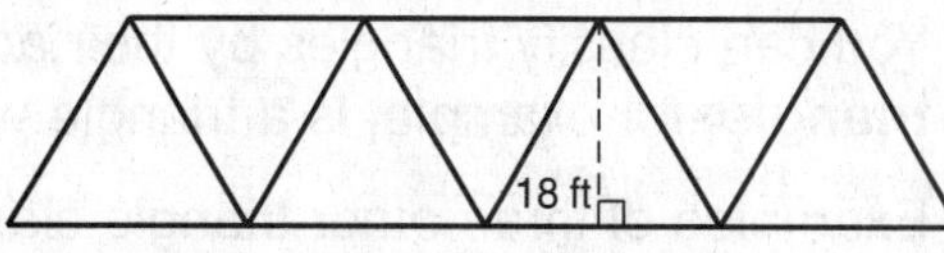

1. Find the total length of steel girder used to make this side of the bridge in feet and inches. Round to the nearest inch. ____________

2. Find the length of wooden planks needed (that is, the span of the bridge) in feet and inches. Round to the nearest inch. ____________

Use the figure for Exercises 3 and 4.

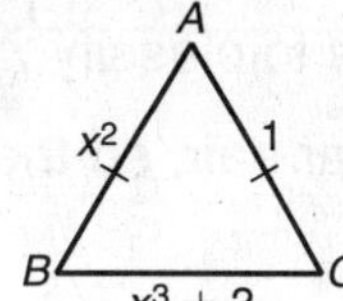

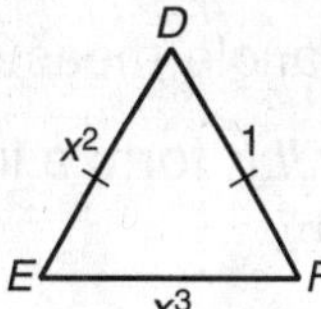

3. Find all possible values of x for each triangle. Explain why the solutions are different.

____

____

____

4. Tell what kind of triangle each must be in all cases.

____

5. Isosceles $\triangle GHI$ has $\overline{GH} \cong \overline{GI}$. $GH = x^2$, $GI = -2x + 15$, and $HI = -x + 4$. Find the side lengths.

____

6. Given that $\overline{AG} \parallel \overline{BF} \parallel \overline{CE}$ and $m\angle A = m\angle D = m\angle G = 60°$, write a paragraph proof proving that all the numbered angles in the figure are congruent. (*Hint:* Mark off each angle as you go.)

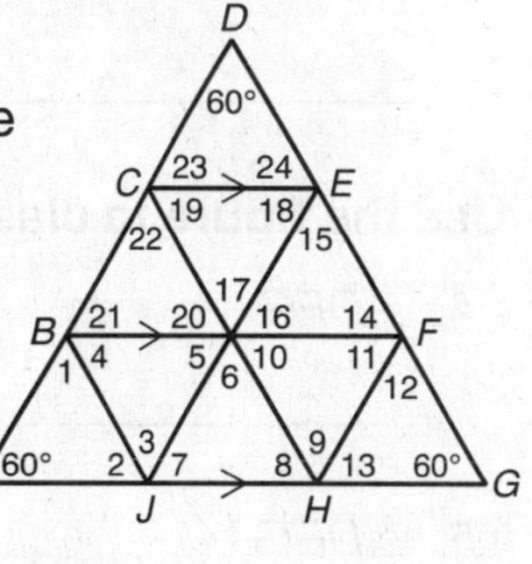

Name ______________________ Date ____________ Class ____________

LESSON 4-1

Reteach

Classifying Triangles

You can classify triangles by their angle measures. An **equiangular triangle,** for example, is a triangle with three congruent angles.

Examples of three other triangle classifications are shown in the table.

A 60° C 60° 60° B

$\angle A \cong \angle B \cong \angle C$

$\triangle ABC$ is equiangular.

Acute Triangle	Right Triangle	Obtuse Triangle
72°, 50°, 58°	53°, 37°	45°, 31°, 104°
all acute angles	one right angle	one obtuse angle

You can use angle measures to classify $\triangle JML$ at right.

$\angle JLM$ and $\angle JLK$ form a linear pair, so they are supplementary.

J 60° M 60° L 120° K

$\angle JKL$ is obtuse so $\triangle JLK$ is an obtuse triangle.

$m\angle JLM + m\angle JLK = 180°$	Def. of supp. ∠s
$m\angle JLM + 120° = 180°$	Substitution
$m\angle JLM = 60°$	Subtract.

Since all the angles in $\triangle JLM$ are congruent, $\triangle JLM$ is an equiangular triangle.

Classify each triangle by its angle measures.

1.

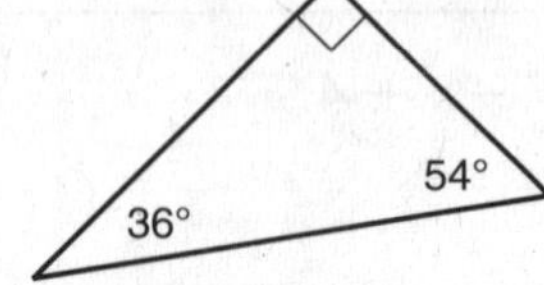

2.

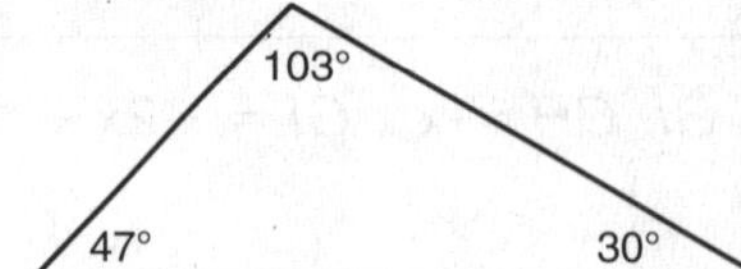

3.

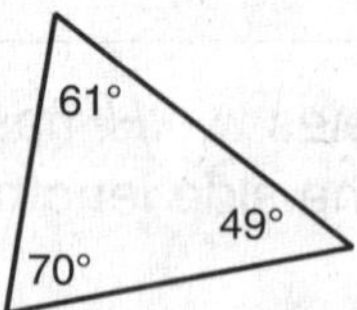

_______________ _______________ _______________

Use the figure to classify each triangle by its angle measures.

4. $\triangle DFG$

5. $\triangle DEG$

____

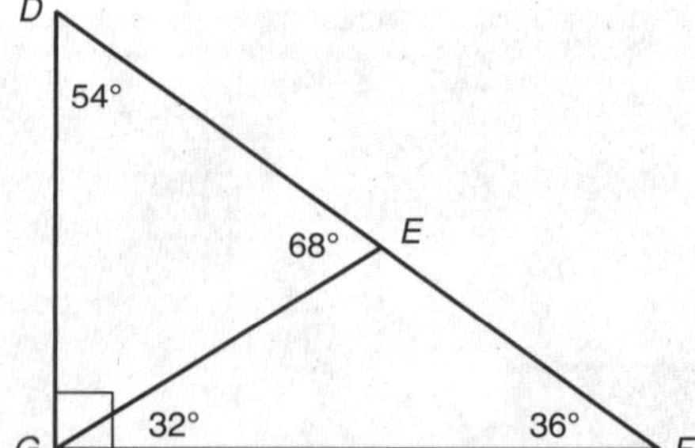

6. $\triangle EFG$

____

Name ______________________ Date ___________ Class ___________

LESSON 4-1

Reteach

Classifying Triangles continued

You can also classify triangles by their side lengths.

Equilateral Triangle	Isosceles Triangle	Scalene Triangle
4, 4, 4 all sides congruent	8, 8 at least two sides congruent	11, 7, 9 no sides congruent

You can use triangle classification to find the side lengths of a triangle.

Step 1 Find the value of x.

$QR = RS$	Def. of $\cong$ segs.
$4x = 3x + 5$	Substitution
$x = 5$	Simplify.

Step 2 Use substitution to find the length of a side.

$4x = 4(5)$	Substitute 5 for x.
$= 20$	Simplify.

Each side length of $\triangle QRS$ is 20.

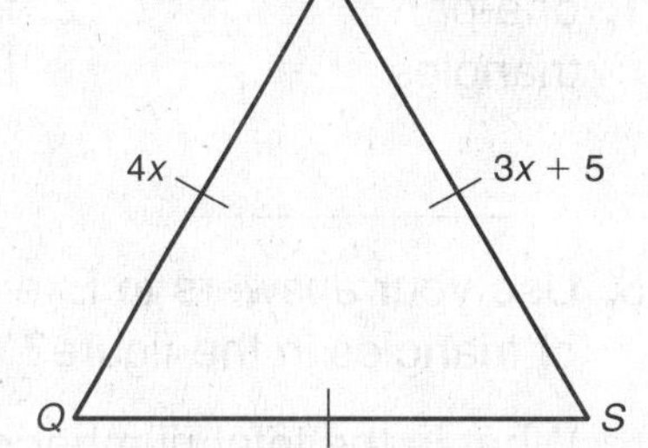

Classify each triangle by its side lengths.

7. $\triangle EGF$

____

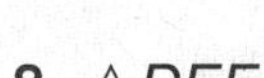

8. $\triangle DEF$

____

9. $\triangle DFG$

____

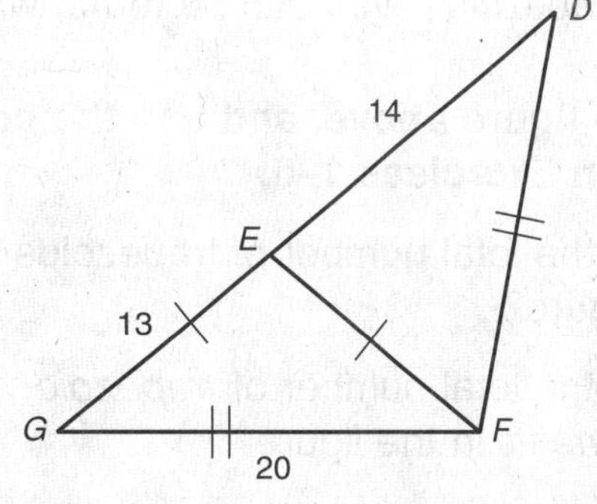

Find the side lengths of each triangle.

10.

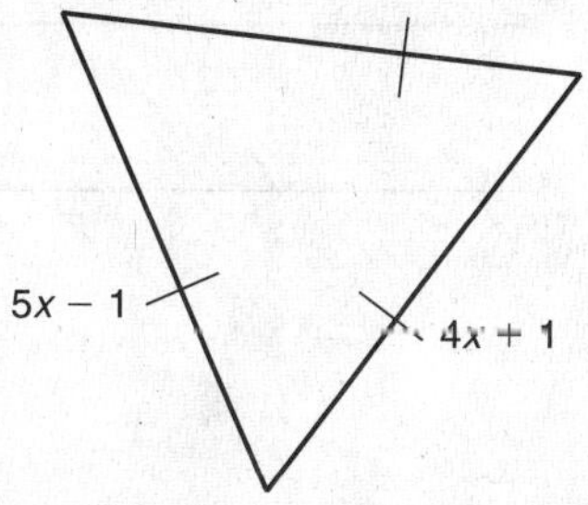

____

11.

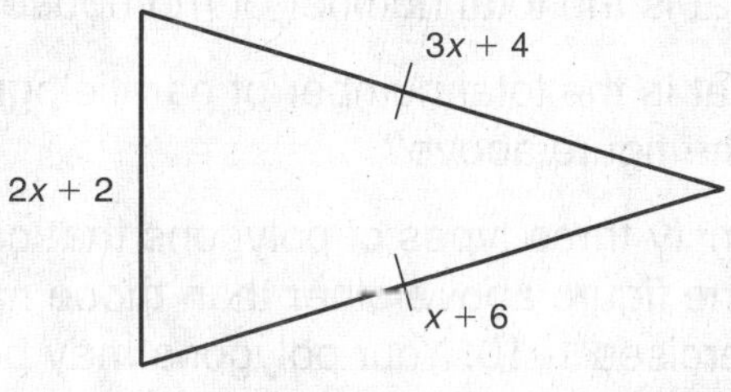

____

Name ______________________ Date ____________ Class ____________

LESSON 4-1

Challenge

A Tantalizing Triangle

How many triangles are in the figure at right? Many people just count the small triangles and decide that there are sixteen in all. However, that answer accounts for only the small "one-part" triangles. Notice that there are actually four types of triangle.

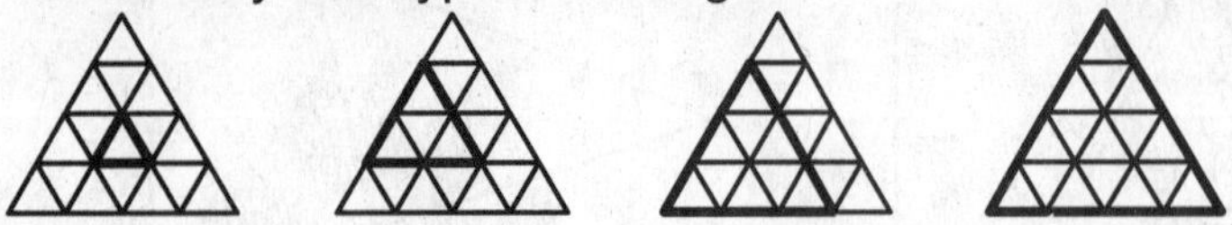

one-part triangles | four-part triangles | nine-part triangles | a sixteen-part triangle

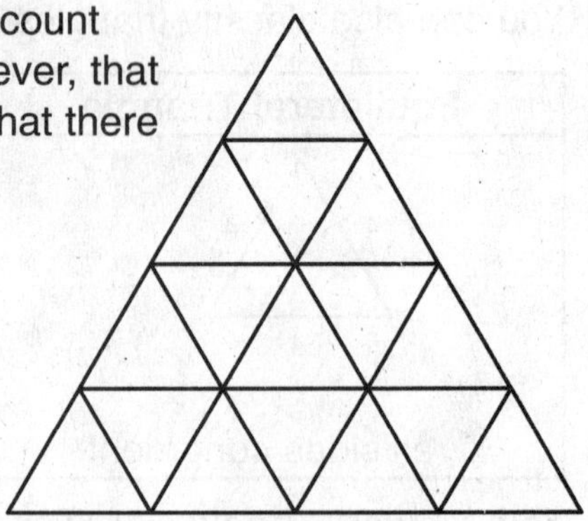

Refer to the figure at right above. How many triangles of each type are there?

1. one-part triangles ____________
2. four-part triangles ____________
3. nine-part triangles ____________
4. sixteen-part triangles ____________

5. Use your answers to Exercises 1–4. What is the total number of triangles in the figure? ____________

6. What is the total number of *triangle midsegments* in the figure? ____________

You may have noticed that there are several other types of geometric shapes within the figure above. For instance, the diagrams at right highlight just three of the many "three-part trapezoids" that can be found within it.

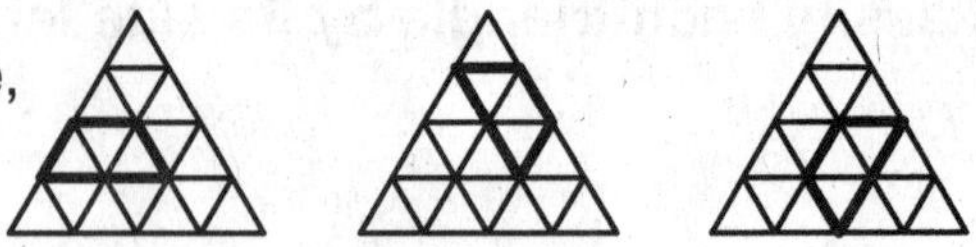

Refer to the figure above, and use the counting techniques developed in Exercises 1–6.

7. What is the total number of trapezoids in the figure? ____________

8. What is the total number of *trapezoid midsegments* in the figure? ____________

9. Assume that all the small triangles are equilateral. What is the total number of rhombuses in the figure above? ____________

10. What is the total number of parallelograms in the figure above? ____________

11. Identify three types of polygons that can be found in the figure above other than those named in Exercises 1–10. Your polygons may be convex or concave. Sketch each polygon in the space at right. Then find the total number of each type of polygon that you identified in the figure. ____________

Name ______________________ Date __________ Class __________

LESSON 4-1

Problem Solving
Classifying Triangles

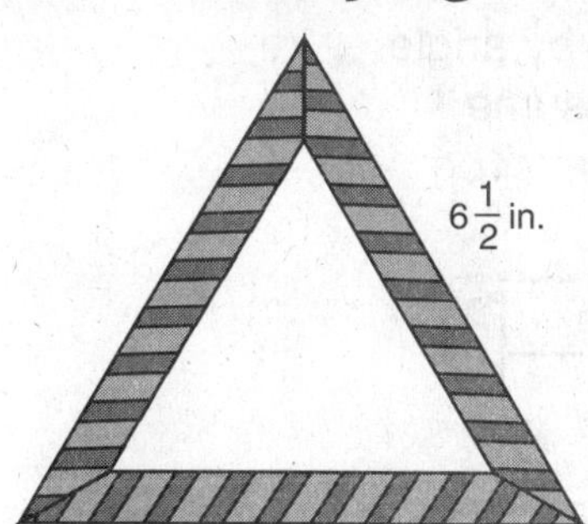

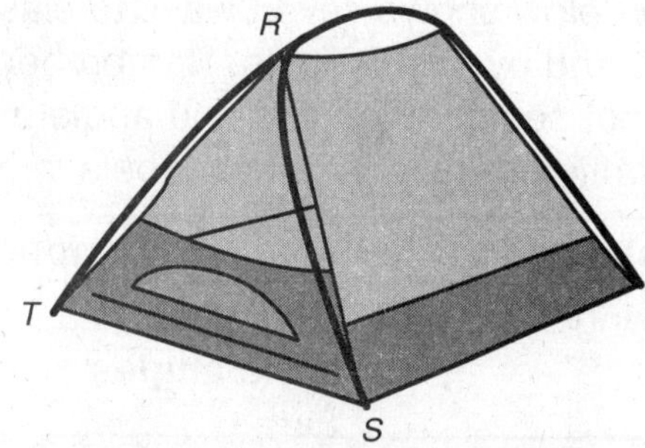

1. Aisha makes triangular picture frames by gluing three pieces of wood together in the shape of an equilateral triangle and covering the wood with ribbon. Each side of a frame is $6\frac{1}{2}$ inches long. How many frames can she cover with 2 yards of ribbon?

____

2. A tent's entrance is in the shape of an isosceles triangle in which $\overline{RT} \cong \overline{RS}$. The length of $\overline{TS}$ is 1.2 times the length of a side. The perimeter of the entrance is 14 feet. Find each side length.

____

Use the figure and the following information for Exercises 3 and 4.

The distance "as the crow flies" between Santa Fe and Phoenix is 609 kilometers. This is 245 kilometers less than twice the distance between Santa Fe and El Paso. Phoenix is 48 kilometers closer to El Paso than it is to Santa Fe.

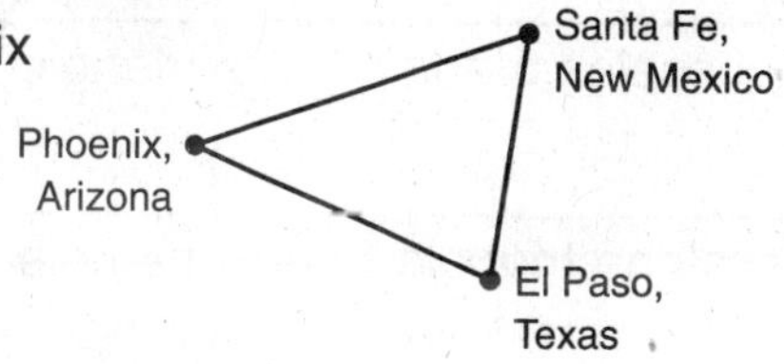

Santa Fe, New Mexico
Phoenix, Arizona
El Paso, Texas

3. What is the distance between each pair of cities?

____

____

4. Classify the triangle that connects the cities by its side lengths. _______________

Choose the best answer.

A *gable,* as shown in the diagram, is the triangular portion of a wall between a sloping roof.

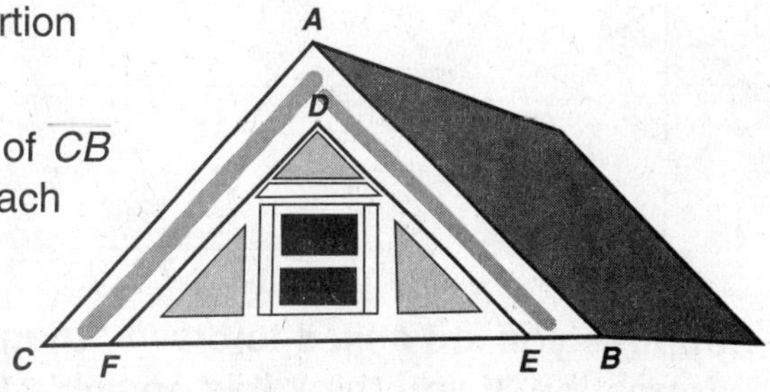

A
D
C F E B

5. Triangle *ABC* is an isosceles triangle. The length of $\overline{CB}$ is 12 feet 4 inches and the congruent sides are each $\frac{3}{4}$ this length. What is the perimeter of $\triangle ABC$?

A 31 ft 4 in. **C** 21 ft 7 in.

B 30 ft 10 in. **D** 18 ft 6 in.

6. In $\triangle DEF$, $\overline{DE}$ and $\overline{DF}$ are each 6 feet 3 inches long. This length is 0.75 times the length of $\overline{FE}$. What is the perimeter of $\triangle DEF$?

F 12 ft 4 in. **H** 17 ft 2 in.

G 14 ft 7 in. **J** 20 ft 10 in.

Name ______________________ Date ____________ Class ____________

LESSON **4-1**

Reading Strategies

Vocabulary Development

The table below shows seven ways to classify different triangles, by angle measures and by side lengths. Remember, you cannot simply assume things about segment lengths and angle measures. Information must be given in writing or by marks and labels in the diagram.

Classification	Description	Example
acute triangle	triangle that has *three* acute **angles**	78° 69° 33°
equiangular triangle	triangle that has *three* congruent acute **angles**	
right triangle	triangle that has *one* right **angle**	
obtuse triangle	triangle that has *one* obtuse **angle**	120°
equilateral triangle	triangle with *three* congruent **sides**	
isosceles triangle	triangle that has *at least two* congruent **sides**	
scalene triangle	triangle that has *no* congruent **sides**	3 5 4

Classify the following triangles by side lengths and angle measures. There will be more than one answer.

1. 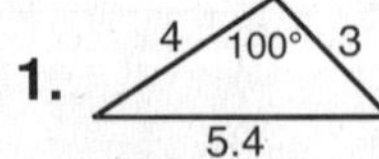

____

2. 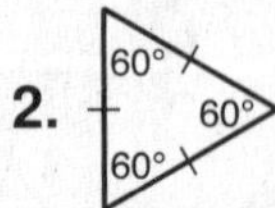

____

3. 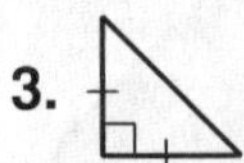
____

Suppose you are asked to draw a triangle by using the given information. If you think it is possible to draw such a triangle, classify the triangle. Otherwise write *no such triangle*.

4. $\triangle ABC$ with $AB = 3$, $BC = 3$, and $CA = 5$ ____________

5. $\triangle XOZ$ with $m\angle X = 92°$, $m\angle O = 92°$, and $m\angle Z = 27°$ ____________

6. $\triangle MNK$ with $m\angle M = 90°$, $m\angle N = 60°$, and $m\angle K = 30°$ ____________

Name ______________________ Date __________ Class __________

LESSON 4-2

Practice A

Angle Relationships in Triangles

Use the figure for Exercises 1–3. Name all the angles that fit the definition of each vocabulary word.

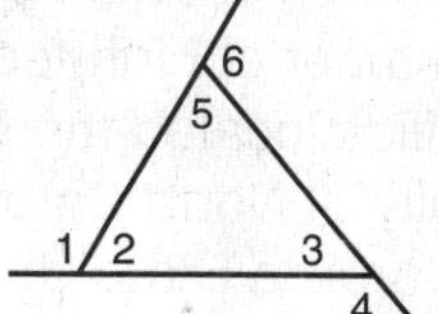

1. exterior angle ______________________

2. remote interior angles to ∠6 ______________________

3. interior angle ______________________

For Exercises 4–7, fill in the blanks to complete each theorem or corollary.

4. The measure of each angle of an ______________________ triangle is 60°.

5. The sum of the angle measures of a triangle is ______________________.

6. The acute angles of a ______________________ triangle are complementary.

7. The measure of an ______________________ of a triangle is equal to the sum of the measures of its remote interior angles.

Find the measure of each angle.

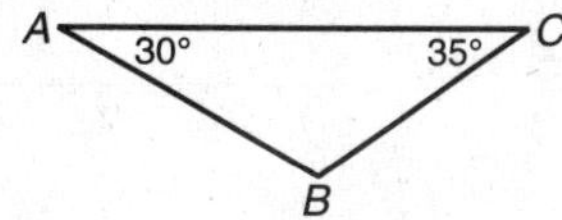

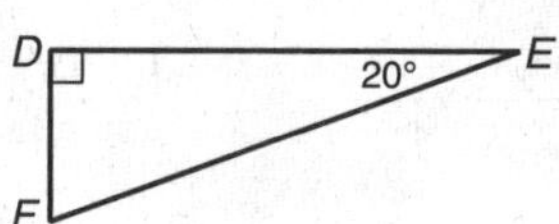

8. m∠*B* ______________________

9. m∠*F* ______________________

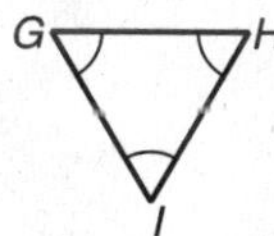

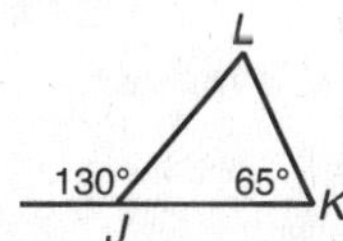

10. m∠*G* ______________________

11. m∠*L* ______________________

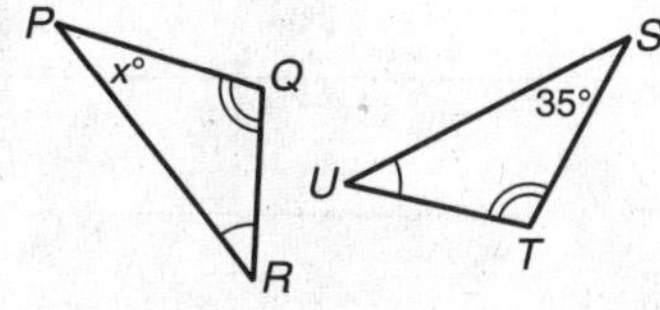

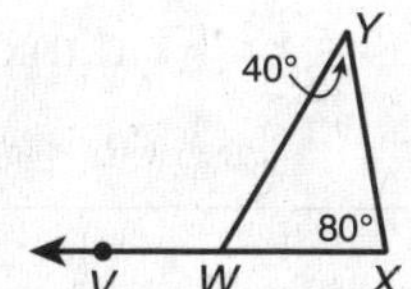

12. m∠*P* ______________________

13. m∠*VWY* ______________________

14. When a person's joint is injured, the person often goes through rehabilitation under the supervision of a doctor or physical therapist to make sure the joint heals well. Rehabilitation involves stretching and exercises. The figure shows a leg bending at the knee during a rehabilitation session. Use what you know about triangles to find the angle measure that the knee is bent from the horizontal (fully extended) position.

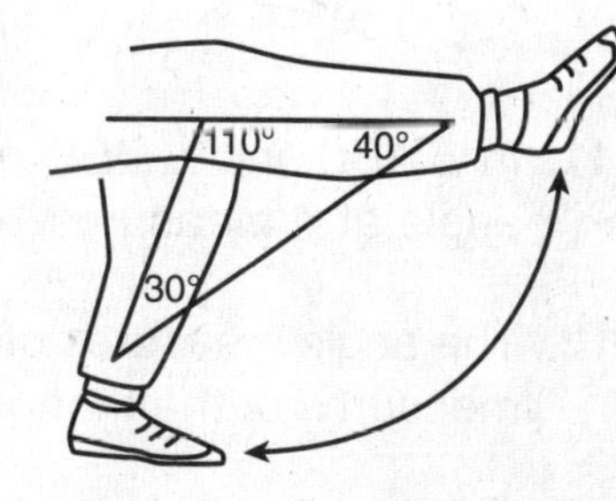

____

Name ________________________________ Date ____________ Class ____________

LESSON 4-2

Practice B

Angle Relationships in Triangles

1. An area in central North Carolina is known as the Research Triangle because of the relatively large number of high-tech companies and research universities located there. Duke University, the University of North Carolina at Chapel Hill, and North Carolina State University are all within this area. The Research Triangle is roughly bounded by the cities of Chapel Hill, Durham, and Raleigh. From Chapel Hill, the angle between Durham and Raleigh measures 54.8°. From Raleigh, the angle between Chapel Hill and Durham measures 24.1°. Find the angle between Chapel Hill and Raleigh from Durham. ____________

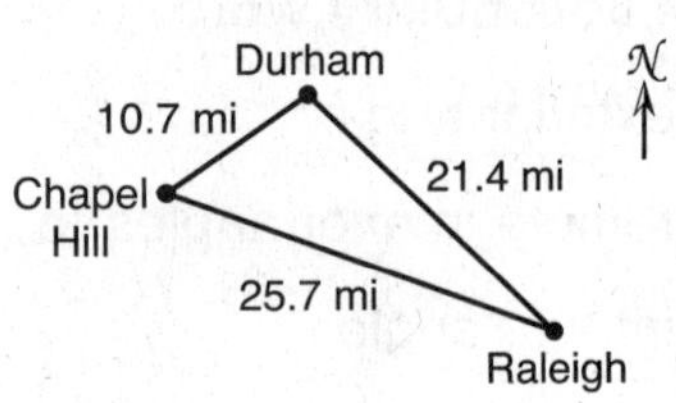

2. The acute angles of right triangle *ABC* are congruent. Find their measures. ____________

The measure of one of the acute angles in a right triangle is given. Find the measure of the other acute angle.

3. 44.9° ____________ 4. $(90 - z)°$ ____________ 5. 0.3° ____________

Find each angle measure.

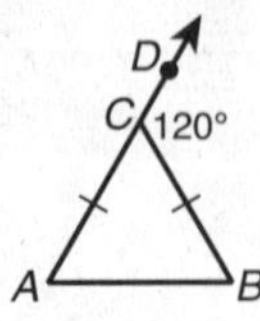

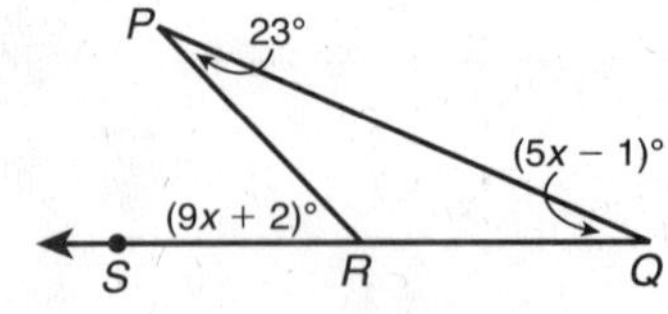

6. m∠*B* ____________ 7. m∠*PRS* ____________

8. In △*LMN*, the measure of an exterior angle at *N* measures 99°. $m\angle L = \frac{1}{3}x°$ and $m\angle M = \frac{2}{3}x°$. Find m∠*L*, m∠*M*, and m∠*LNM*. ____________

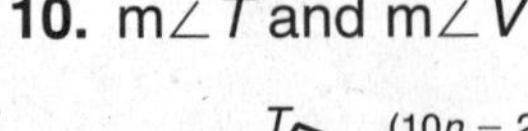

9. m∠*E* and m∠*G* ____________ 10. m∠*T* and m∠*V* ____________

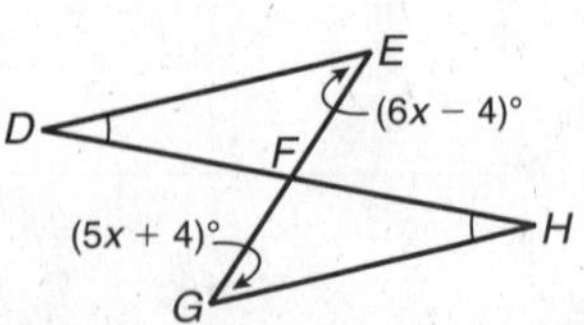

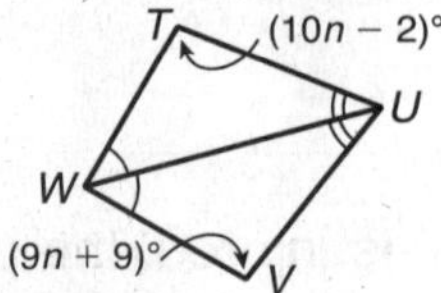

11. In △*ABC* and △*DEF*, m∠*A* = m∠*D* and m∠*B* = m∠*E*. Find m∠*F* if an exterior angle at *A* measures 107°, $m\angle B = (5x + 2)°$, and $m\angle C = (5x + 5)°$. ____________

12. The angle measures of a triangle are in the ratio 3 : 4 : 3. Find the angle measures of the triangle. ____________

Name ______________________ Date __________ Class __________

LESSON 4-2

Practice C

Angle Relationships in Triangles

1. Write a two-column proof that the sum of the angle measures of a quadrilateral is 360°. Begin by drawing quadrilateral *ABCD*. (*Hint:* You will have to draw one auxiliary line.)

2. Use this figure to write a flowchart proof that the sum of the measures of the exterior angles of a triangle, one at each vertex, is 360°.

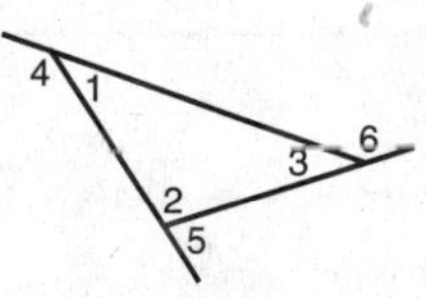

3. Find the sum of the exterior angles, one at each vertex, of a quadrilateral. ____________

4. Use the techniques you developed in Exercises 1–3 to find the sums of the measures of the interior angles and of the exterior angles, one at each vertex, of a pentagon.

____

5. A landscape artist plans to draw a pair of mountains. He wants his drawing to be reasonably accurate, so he takes some measurements and draws this figure. Find *x*, *y*, and *z*. ______________________

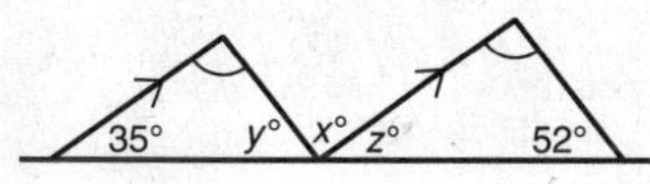

Name ______________________ Date __________ Class __________

LESSON 4-2

Reteach

Angle Relationships in Triangles

According to the **Triangle Sum Theorem,** the sum of the angle measures of a triangle is 180°.

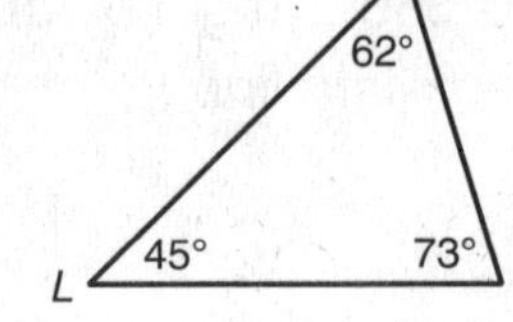

$m\angle J + m\angle K + m\angle L = 62 + 73 + 45$

$= 180°$

The **corollary** below follows directly from the Triangle Sum Theorem.

Corollary	Example
The acute angles of a right triangle are complementary.	$m\angle C = 90 - 39$ $= 51°$ $m\angle C + m\angle E = 90°$

Use the figure for Exercises 1 and 2.

1. Find $m\angle ABC$.

____

2. Find $m\angle CAD$.

____

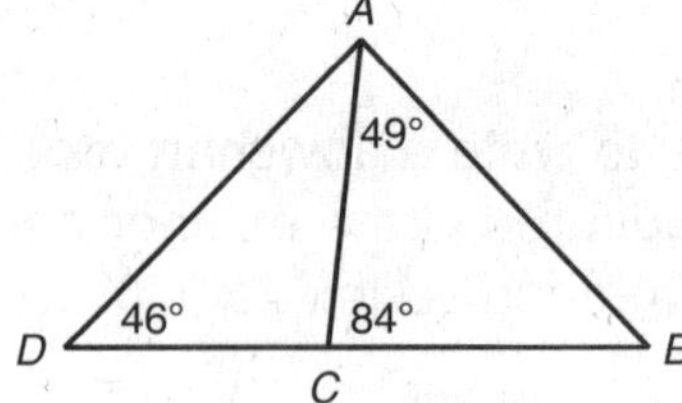

Use $\triangle RST$ for Exercises 3 and 4.

3. What is the value of x?

____

4. What is the measure of each angle?

____

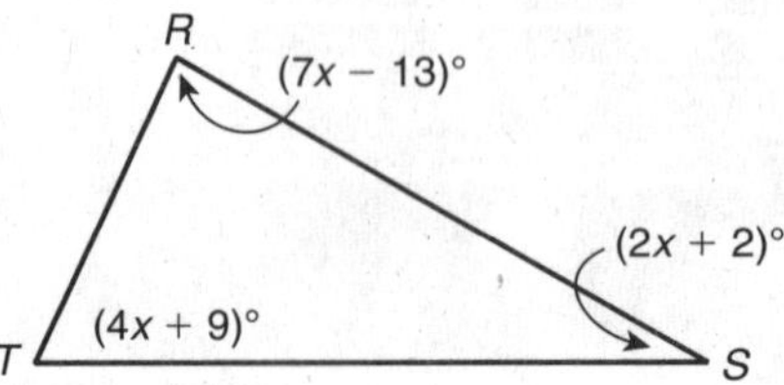

What is the measure of each angle?

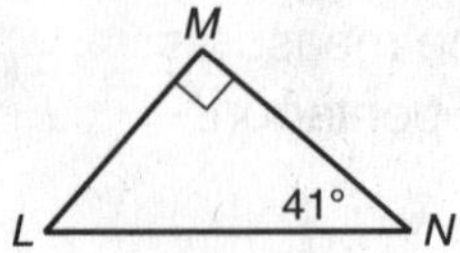

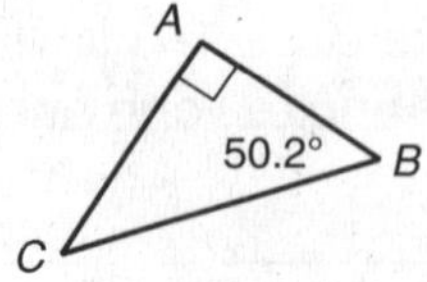

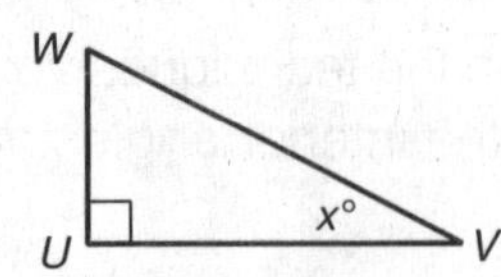

5. $\angle L$ ______________

6. $\angle C$ ______________

7. $\angle W$ ______________

Name ______________________ Date __________ Class __________

LESSON 4-2

Reteach

Angle Relationships in Triangles continued

An **exterior angle** of a triangle is formed by one side of the triangle and the extension of an adjacent side.

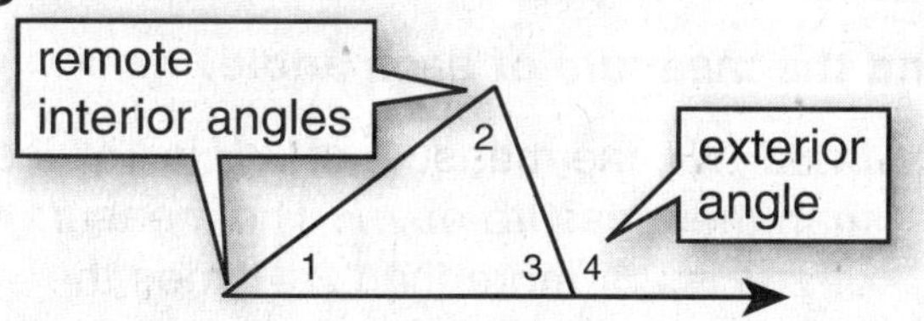

∠1 and ∠2 are the remote interior angles of ∠4 because they are not adjacent to ∠4.

Exterior Angle Theorem

The measure of an exterior angle of a triangle is equal to the sum of the measures of its remote interior angles.

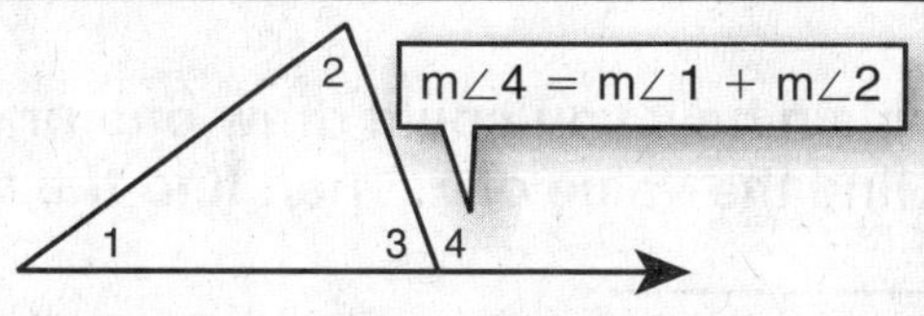

Third Angles Theorem

If two angles of one triangle are congruent to two angles of another triangle, then the third pair of angles are congruent.

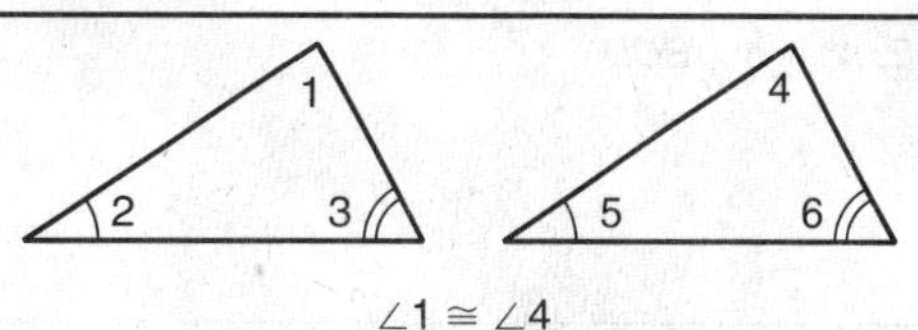

Find each angle measure.

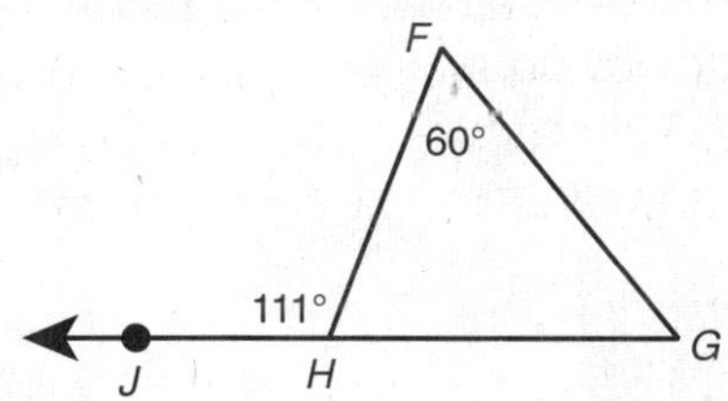

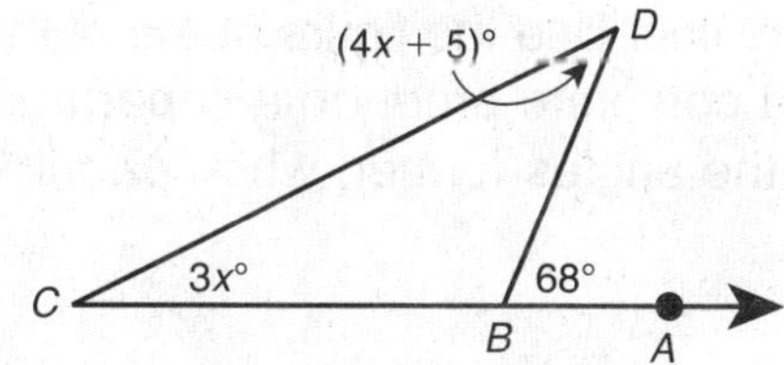

8. m∠*G*

____

9. m∠*D*

____

Find each angle measure.

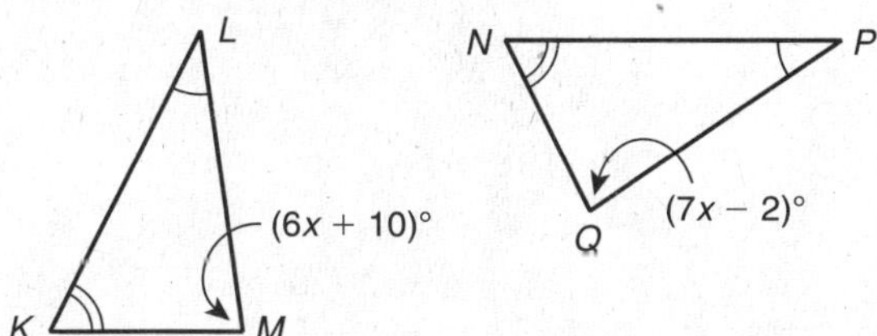

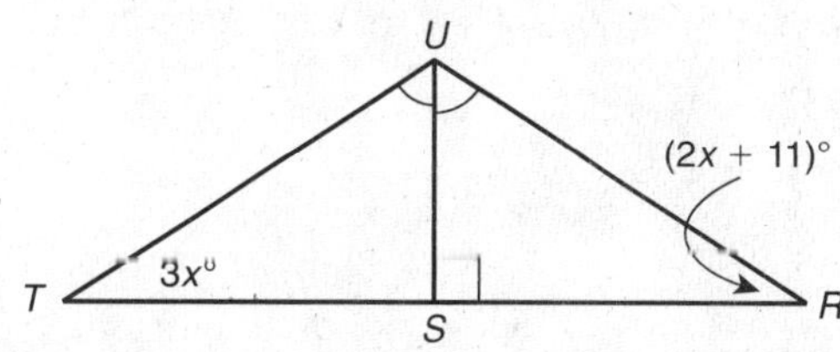

10. m∠*M* and m∠*Q*

____

11. m∠*T* and m∠*R*

____

Name ________________________________ Date ___________ Class ___________

LESSON 4-2

Challenge

Analyzing Verbal Descriptions and Using Auxiliary Figures

Find the measure of each angle.

1. In $\triangle FGH$, the measure of $\angle H$ is 14° less than the measure of $\angle F$. The measure of $\angle G$ is 25° more than $2\frac{1}{3}$ times the measure of $\angle F$. Find $m\angle F$.

____

2. In $\triangle WXY$, the measure of $\angle X$ is 4° less than twice the measure of $\angle W$. The measure of $\angle Y$ is 24° less than $3\frac{1}{2}$ times $m\angle W$. Find $m\angle Y$.

____

$\overline{BA} \parallel \overline{DE}$. Explain how you could draw one or more auxiliary figures to help you find the value of *x*. Then find the value of *x*. Explain.

3. 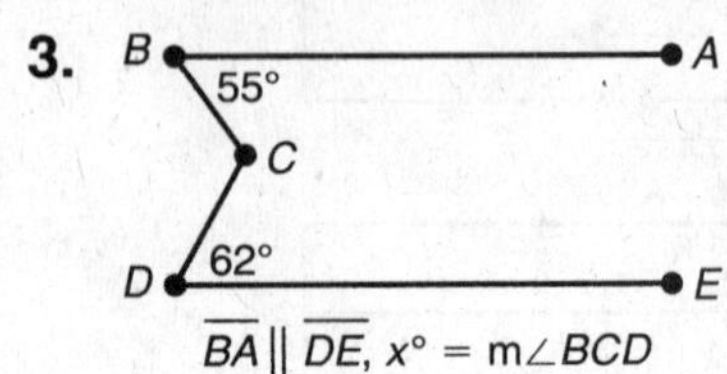

____

____

____

4. It is possible to prove the Exterior Angle Theorem by drawing an auxiliary line. Use the figure at the right to show how this might be done. Write a complete proof on a separate sheet of paper. (*Hint:* Think about the angles formed when parallel lines are cut by a transversal.)

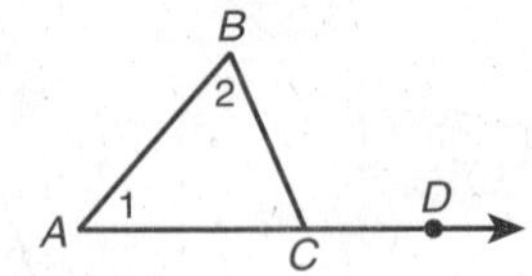

Name ______________________ Date __________ Class __________

LESSON 4-2

Problem Solving

Angle Relationships in Triangles

1. The locations of three food stands on a fair's midway are shown. What is the measure of the angle labeled $x°$?

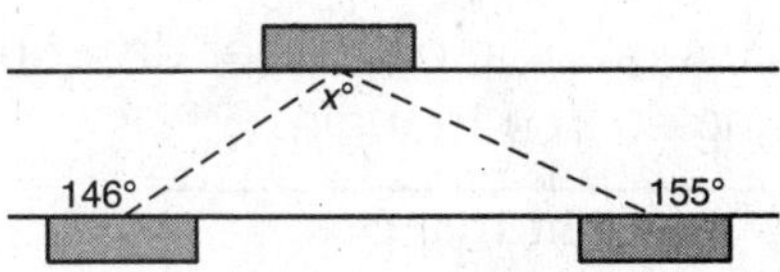

____

2. A large triangular piece of plywood is to be painted to look like a mountain for the spring musical. The angles at the base of the plywood measure 76° and 45°. What is the measure of the top angle that represents the mountain peak?

____

Use the figure of the banner for Exercises 3 and 4.

3. What is the value of n?

____

4. What is the measure of each angle in the banner?

____

Use the figure of the athlete pole vaulting for Exercises 5 and 6.

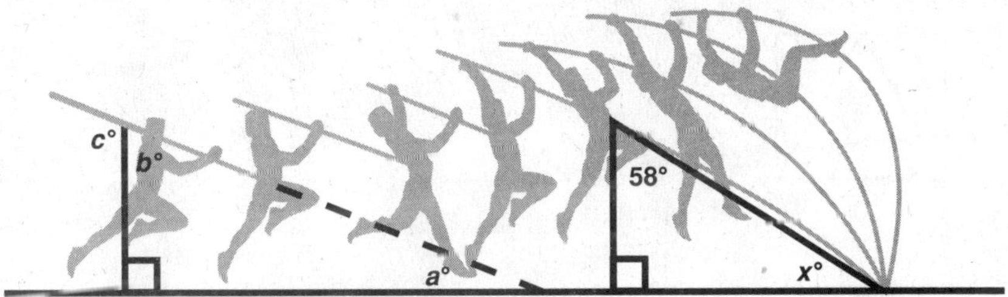

5. What is $x°$, the measure of the angle that the pole makes when it first touches the ground?

____

6. At takeoff, $a° = 23°$. What is $c°$, the measure of the angle the pole makes with the athlete's body?

____

The figure shows a path through a garden. Choose the best answer.

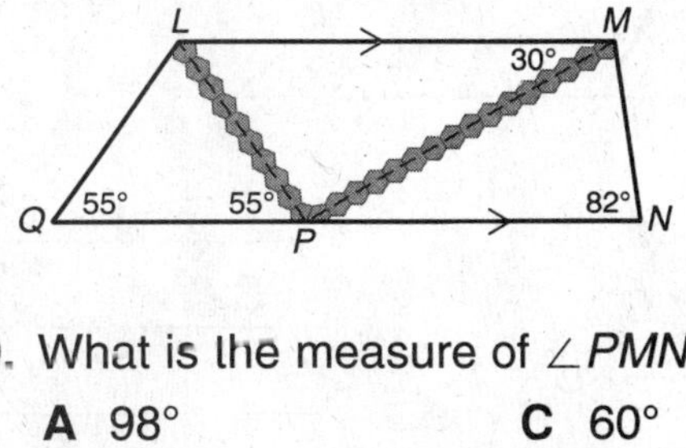

7. What is the measure of $\angle QLP$?

   A 20° **C** 110°

   B 70° **D** 125°

8. What is the measure of $\angle LPM$?

   F 85° **H** 95°

   G 90° **J** 125°

9. What is the measure of $\angle PMN$?

   A 98° **C** 60°

   B 68° **D** 55°

Name ______________________ Date __________ Class __________

LESSON 4-2

Reading Strategies

Graphic Organizer

This graphic organizer describes the relationships of interior and exterior angles in a triangle.

Right Triangles

The *acute angles* of a right triangle are *complementary*.

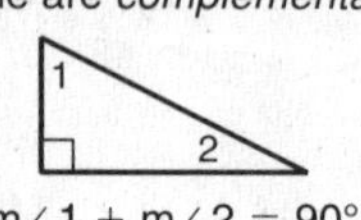

m∠1 + m∠2 = 90°

Equiangular Triangles

The measure of *each angle* in an equiangular triangle is *60°*.

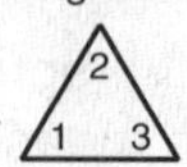

m∠1 = 60°, m∠2 = 60°, m∠3 = 60°

Triangle Sum Theorem

The sum of the angle measures of a triangle is 180°.

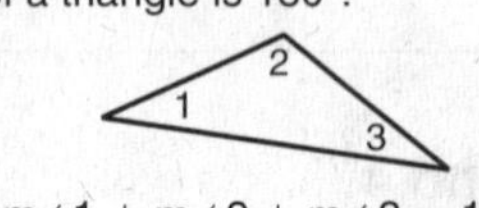

m∠1 + m∠2 + m∠3 = 180°

Exterior Angle Theorem

The measure of an *exterior angle* of a triangle is equal to the *sum of the measures of its remote interior angles*.

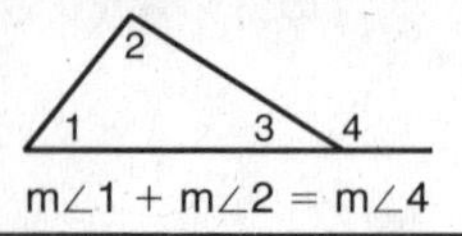

m∠1 + m∠2 = m∠4

Third Angle Theorem

If two angles of one triangle are congruent to two angles of another triangle, then *the third pair of angles are congruent*.

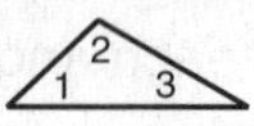

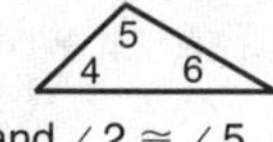

If ∠1 ≅ ∠4 and ∠2 ≅ ∠5, then ∠3 ≅ ∠6.

Use the given information to find the measures of the angles.

∠*S* and ∠*Q* are right angles.
m∠*QPR* = 30°
△*TRP* is equiangular.

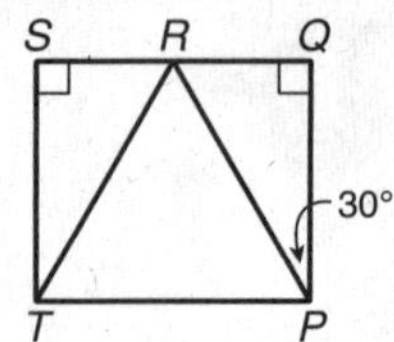

1. Find m∠*QRP*. ______________________
2. Find m∠*TRP*. ______________________
3. Find m∠*RTS*. ______________________

Use the figure for Exercises 4–7.

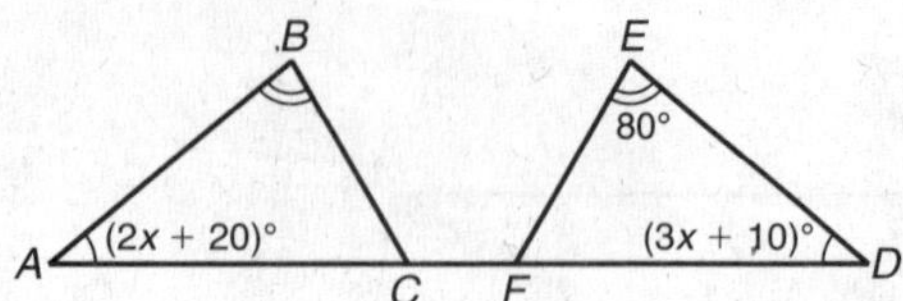

4. Find m∠*A*. ______________________
5. Find m∠*B*. ______________________
6. Find m∠*BCF*. ______________________
7. Find m∠*EFD*. ______________________

Name ______________________ Date __________ Class __________

LESSON 4-3

Practice A
Congruent Triangles

Fill in the blanks to complete each definition.

1. ____________________ angles and ____________________ sides are in the same position in polygons with an equal number of sides.

2. Two polygons are ____________________ polygons if and only if their corresponding angles and sides are congruent.

Refer to the figure of $\triangle GHI$ and $\triangle JKL$ for Exercises 3 and 4.

3. Name the three pairs of corresponding sides.

____

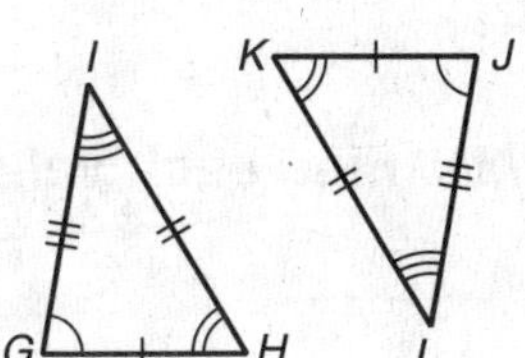

4. Name the three pairs of corresponding angles.

____

Find the value of *x*.

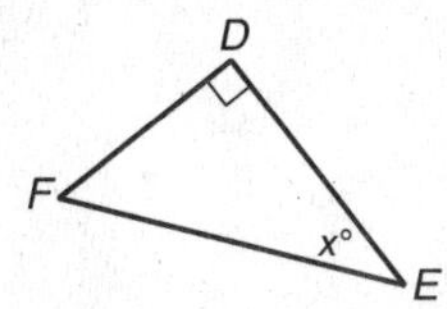

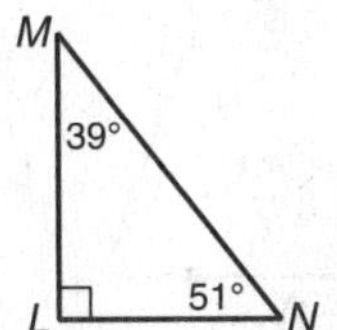

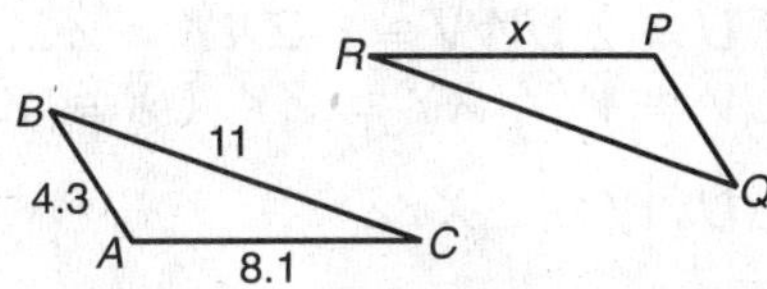

5. Given: $\triangle DEF \cong \triangle LMN$ ______________ **6. Given:** $\triangle ABC \cong \triangle PQR$ ______________

7. Etienne flies a kite. When the kite is flying well, the tail sticks out straight so the indicated angles at *V* are congruent. Use the phrases from the word bank to complete this two-column proof.

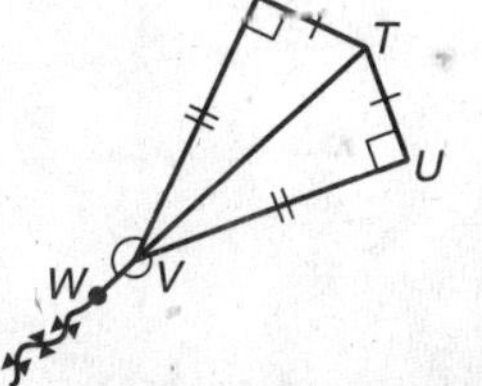

$\angle S \cong \angle U$
Third ∠s Thm.
Given
$\angle SVT \cong \angle UVT$

Given: $\angle S$ and $\angle U$ are right angles.
$\angle SVW \cong \angle UVW$, $\overline{SV} \cong \overline{UV}$, $\overline{ST} \cong \overline{UT}$

Prove: $\triangle STV \cong \triangle UTV$

Proof:

Statements	Reasons
1. $\overline{SV} \cong \overline{UV}$, $\overline{ST} \cong \overline{UT}$	1. Given
2. $\overline{TV} \cong \overline{TV}$	2. Reflex. Prop. of $\cong$
3. $\angle S$ and $\angle U$ are right angles.	3. **a.** ____________________
4. **b.** ____________________	4. Rt. $\angle \cong$ Thm.
5. $\angle SVW \cong \angle UVW$	5. Given
6. $\angle SVW$ and $\angle SVT$ are supplementary, $\angle UVW$ are $\angle UVT$ are supplementary.	6. Lin. Pair Thm.
7. **c.** ____________________	7. $\cong$ Suppls. Thm.
8. $\angle STV \cong \angle UTV$	8. **d.** ____________________
9. $\triangle STV \cong \triangle UTV$	9. Def. of $\cong$ $\triangle$s

Name ______________________________ Date ___________ Class ___________

LESSON 4-3

Practice B

Congruent Triangles

In baseball, home plate is a pentagon. Pentagon *ABCDE* is a diagram of a regulation home plate. The baseball rules are very specific about the exact dimensions of this pentagon so that every home plate is congruent to every other home plate. If pentagon *PQRST* is another home plate, identify each congruent corresponding part.

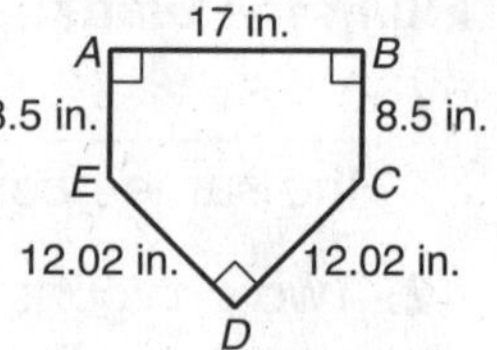

1. $\angle S \cong$ ________ **2.** $\angle B \cong$ ________ **3.** $\overline{EA} \cong$ ________

4. $\angle E \cong$ ________ **5.** $\overline{PQ} \cong$ ________ **6.** $\overline{TS} \cong$ ________

Given: $\triangle DEF \cong \triangle LMN$. Find each value.

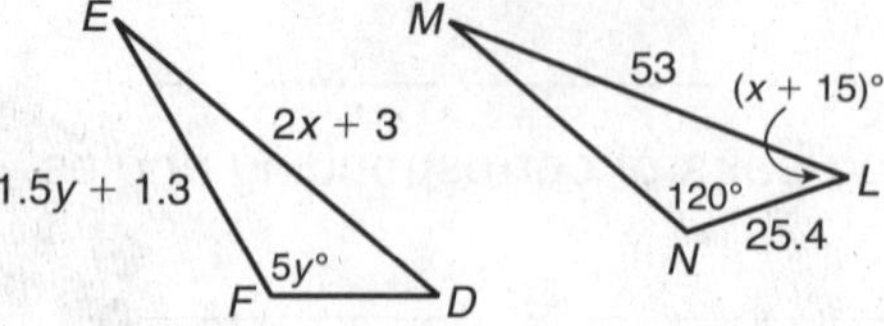

7. $m\angle L =$ ______________

8. $EF =$ ______________

9. Write a two-column proof.

Given: $\angle U \cong \angle UWV \cong \angle ZXY \cong \angle Z$, $\overline{UV} \cong \overline{WV}$, $\overline{XY} \cong \overline{ZY}$, $\overline{UX} \cong \angle \overline{WZ}$

Prove: $\triangle UVW \cong \triangle XYZ$

Proof:

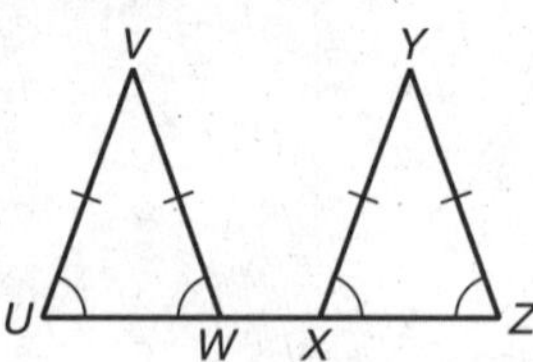

10. Given: $\triangle CDE \cong \triangle HIJ$, $DE = 9x$, and $IJ = 7x + 3$. Find x and DE.

____

11. Given: $\triangle CDE \cong \triangle HIJ$, $m\angle D = (5y + 1)°$, and $m\angle I = (6y - 25)°$. Find y and $m\angle D$.

____

Name ______________________ Date __________ Class __________

LESSON 4-3

Practice C

Congruent Triangles

Mr. X is an inventive person. He takes pleasure in drawing a triangle and seeing if another person can recreate his drawing from piecemeal information. For each exercise, draw a diagram to support your answer. (*Hint:* Begin each exercise by drawing a triangle. Measure the parts of your triangle that Mr. X gives you and try to draw a different triangle with those parts. If the two triangles are congruent, you drew Mr. X's triangle.)

1. If Mr. X gives you the measures of the sides of a triangle, could you be sure you would draw Mr. X's triangle?

   ____

2. If Mr. X gives you the measures of the angles of a triangle, could you be sure you would draw Mr. X's triangle?

   ____

3. If Mr. X gives you the measures of one angle and of both sides of that angle, could you be sure you would draw Mr. X's triangle?

   ____

4. If Mr. X gives you the measures of one side and both angles that share that side, could you be sure you would draw Mr. X's triangle?

   ____

5. If Mr. X gives you the measures of one angle, one adjacent side, and the side opposite the angle, could you be sure you would draw Mr. X's triangle? (*Hint:* Start with an angle less than 45° and a long adjacent side.)

   ____

Name ______________________ Date __________ Class __________

LESSON 4-3

Reteach

Congruent Triangles

Triangles are **congruent** if they have the same size and shape. Their **corresponding parts,** the angles and sides that are in the same positions, are congruent.

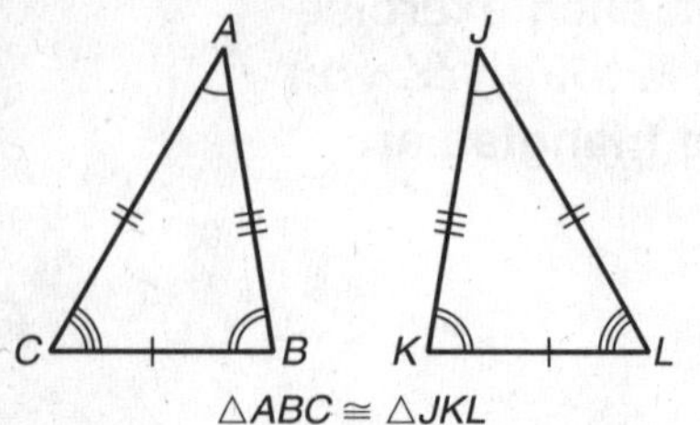

$\triangle ABC \cong \triangle JKL$

Corresponding Parts	
Congruent Angles	**Congruent Sides**
$\angle A \cong \angle J$	$\overline{AB} \cong \overline{JK}$
$\angle B \cong \angle K$	$\overline{BC} \cong \overline{KL}$
$\angle C \cong \angle L$	$\overline{CA} \cong \overline{LJ}$

To identify corresponding parts of congruent triangles, look at the order of the vertices in the congruence statement such as $\triangle ABC \cong \triangle JKL$.

Given: $\triangle XYZ \cong \triangle NPQ$. Identify the congruent corresponding parts.

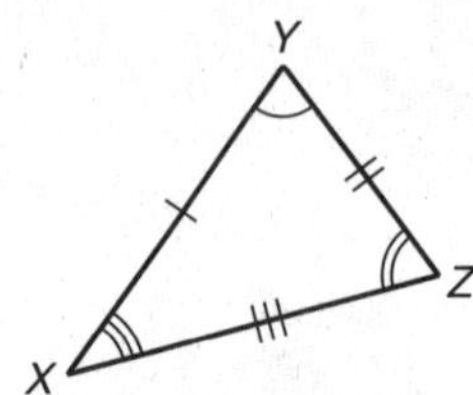

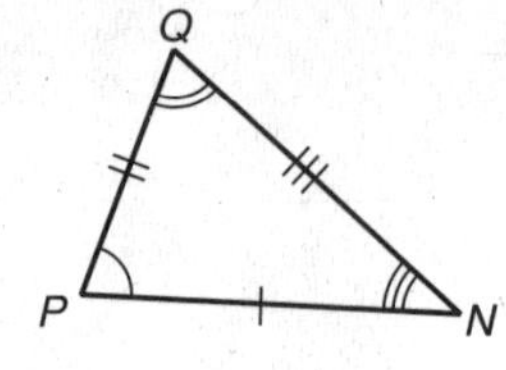

1. $\angle Z \cong$ ______________________

2. $\overline{YZ} \cong$ ______________________

3. $\angle P \cong$ ______________________

4. $\angle X \cong$ ______________________

5. $\overline{NQ} \cong$ ______________________

6. $\overline{PN} \cong$ ______________________

Given: $\triangle EFG \cong \triangle RST$. Find each value below.

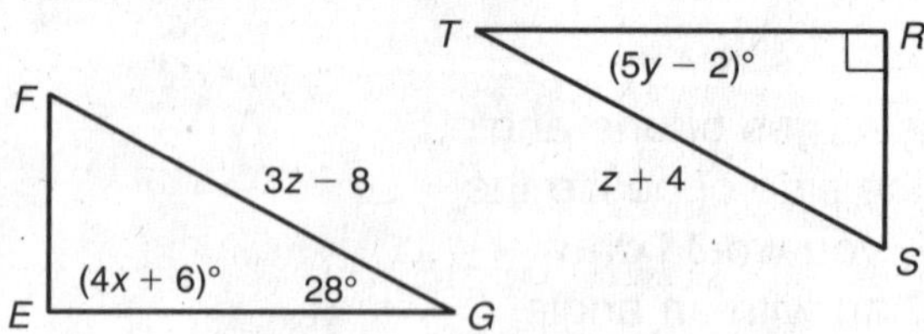

7. $x =$ ______________________

8. $y =$ ______________________

9. $m\angle F =$ ______________________

10. $ST =$ ______________________

Name ________________________________ Date ____________ Class ____________

LESSON 4-3

Reteach

Congruent Triangles continued

You can prove triangles congruent by using the definition of congruence.

Given: $\angle D$ and $\angle B$ are right angles.

$\angle DCE \cong \angle BCA$

C is the midpoint of $\overline{DB}$.

$\overline{ED} \cong \overline{AB}$, $\overline{EC} \cong \overline{AC}$

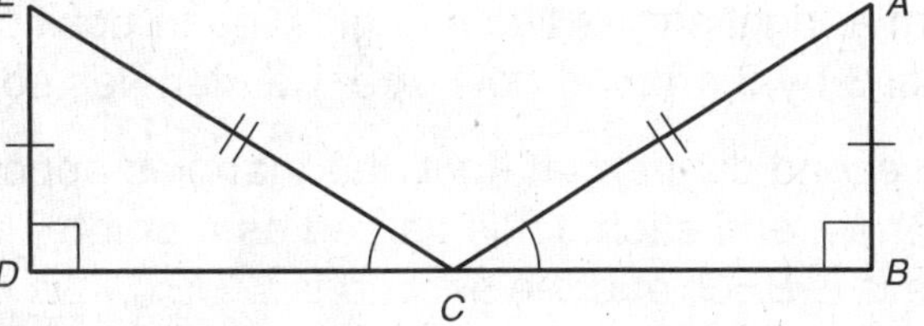

Prove: $\triangle EDC \cong \triangle ABC$

Proof:

Statements	Reasons
1. $\angle D$ and $\angle B$ are rt. $\angle$s.	1. Given
2. $\angle D \cong \angle B$	2. Rt. $\angle \cong$ Thm.
3. $\angle DCE \cong \angle BCA$	3. Given
4. $\angle E \cong \angle A$	4. Third $\angle$s Thm.
5. C is the midpoint of $\overline{DB}$.	5. Given
6. $\overline{DC} \cong \overline{BC}$	6. Def. of mdpt.
7. $\overline{ED} \cong \overline{AB}$, $\overline{EC} \cong \overline{AC}$	7. Given
8. $\triangle EDC \cong \triangle ABC$	8. Def. of $\cong$ $\triangle$s

11. Complete the proof.

Given: $\angle Q \cong \angle R$

P is the midpoint of $\overline{QR}$.

$\overline{NQ} \cong \overline{SR}$, $\overline{NP} \cong \overline{SP}$

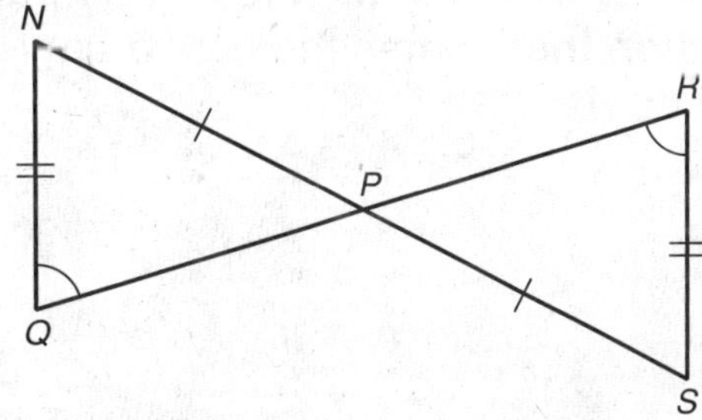

Prove: $\triangle NPQ \cong \triangle SPR$

Proof:

Statements	Reasons
1. $\angle Q \cong \angle R$	1. Given
2. $\angle NPQ \cong \angle SPR$	2. **a.** ____________
3. $\angle N \cong \angle S$	3. **b.** ____________
4. P is the midpoint of $\overline{QR}$.	4. **c.** ____________
5. **d.** ____________	5. Def. of mdpt.
6. $\overline{NQ} \cong \overline{SR}$, $\overline{NP} \cong \overline{SP}$	6. **e.** ____________
7. $\triangle NPQ \cong \triangle SPR$	7. **f.** ____________

Name ____________________ Date __________ Class __________

LESSON 4-3

Challenge

Congruence and Transformations on an Array

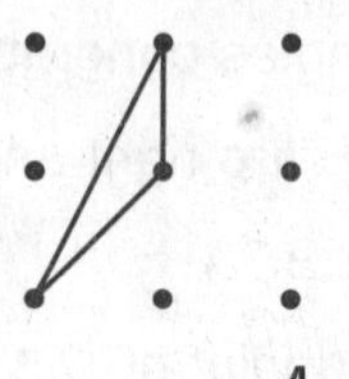

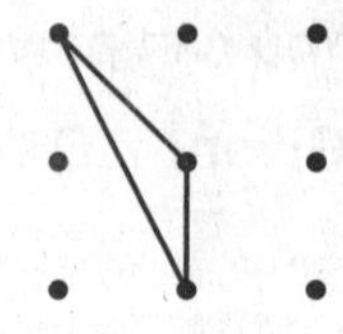

When two geometric figures are congruent, each is the image of the other under a rigid transformation. The first diagram at right shows two triangles, each positioned on an identical 3-by-3 array of dots. Are the triangles congruent?

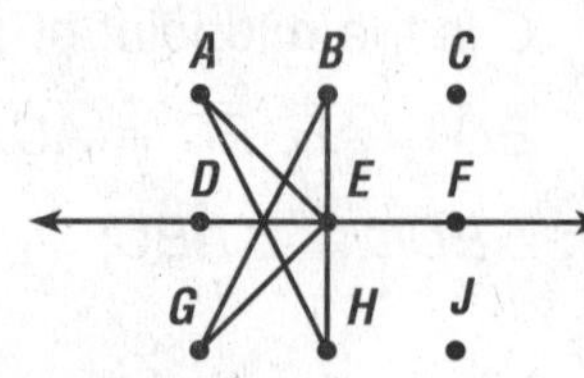

In the second diagram at right, the triangles appear on the same array, and each dot is named as a point. The first triangle is $\triangle BEG$, and the second is $\triangle HEA$. It is clear that $\triangle HEA$ is a reflection of $\triangle BEG$ across a line, $\overleftrightarrow{DF}$. So $\triangle HEA$ is congruent to $\triangle BEG$.

Each triangle is congruent to $\triangle BEG$. Identify the transformation that relates the triangle to $\triangle BEG$.

1. A B C / D E F / G H J

   ____

   ____

2. A B C / D E F / G H J

   ____

   ____

3. A B C / D E F / G H J

   ____

   ____

4. A B C / D E F / G H J

   ____

   ____

On each grid, sketch a triangle congruent to $\triangle BEG$ different from those given above. Use only the labeled points as vertices. Name the transformation that relates the new triangle to $\triangle BEG$.

5. A B C / D E F / G H J

   ____

   ____

6. A B C / D E F / G H J

   ____

   ____

7. A B C / D E F / G H J

   ____

   ____

8. A B C / D E F / G H J

   ____

   ____

9. Refer to the 3-by-3 grids in Exercises 1–8. Using the labeled points as vertices, how many triangles congruent to $\triangle BEG$ are there in all? List them.

   ____

10. Refer to the 3-by-3 grids in Exercises 1–8. Using the labeled points as vertices, how many triangles can be formed on each grid? List them on a separate sheet of paper, dividing the list into groups of congruent triangles.

    ____

Name ______________________ Date __________ Class __________

LESSON **4-3**

Problem Solving

Congruent Triangles

Use the diagram of the fence for Exercises 1 and 2.

$\triangle RQW \cong \triangle TVW$

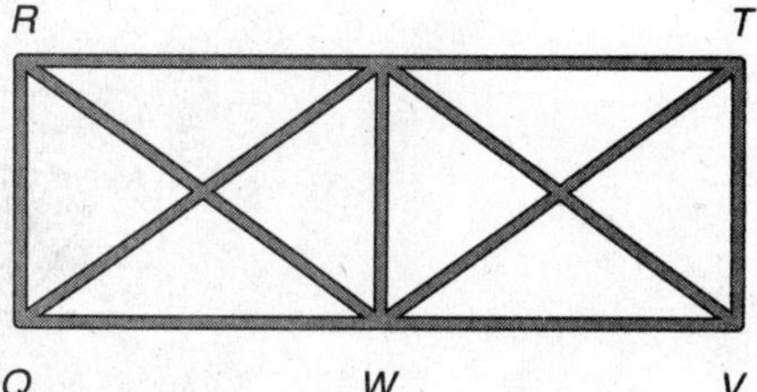

1. If $m\angle RWQ = 36°$ and $m\angle TWV = (2x + 5)°$, what is the value of x?

____

2. If $RW = (3y - 1)$ feet and $TW = (y + 5)$ feet, what is the length of $\overline{RW}$?

____

Use the diagram of a section of the Bank of China Tower for Exercises 3 and 4.

$\triangle JKL \cong \triangle LHJ$

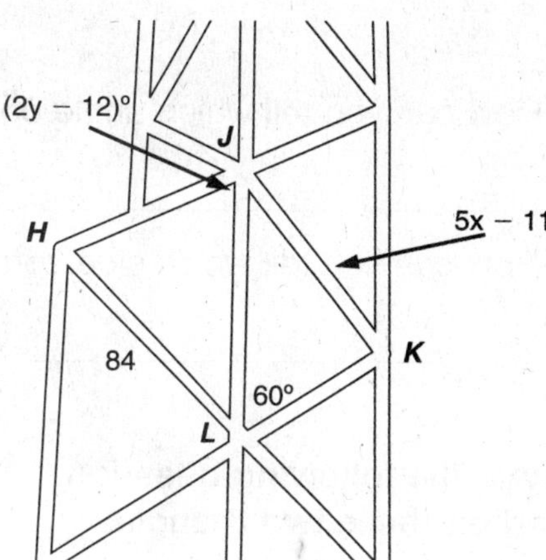

3. What is the value of x?

____

4. Find $m\angle JHL$.

____

Choose the best answer.

5. Chairs with triangular seats were popular in the Middle Ages. Suppose a chair has a seat that is an isosceles triangle and the congruent sides measure $1\frac{1}{2}$ feet. A second chair has a triangular seat with a perimeter of $5\frac{1}{10}$ feet, and it is congruent to the first seat. What is a side length of the second seat?

   A $1\frac{4}{5}$ ft **C** 3 ft

   B $2\frac{1}{10}$ ft **D** $3\frac{3}{5}$ ft

Use the diagram for Exercises 6 and 7.

6. C is the midpoint of $\overline{EB}$ and $\overline{AD}$. What additional information would allow you to prove $\triangle ABC \cong \triangle DEC$ by the definition of congruent triangles?

   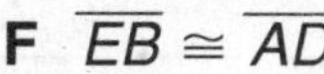

   F $\overline{EB} \cong \overline{AD}$ **H** $\angle ECD \cong \angle ACB$

   G $\overline{DE} \cong \overline{AB}$ **J** $\angle A \cong \angle D$, $\angle B \cong \angle E$

7. If $\triangle ABC \cong \triangle DEC$, $ED = 4y + 2$, and $AB = 6y - 4$, what is the length of $\overline{AB}$?

   A 3 **C** 14

   B 12 **D** 18

Name ______________________ Date ____________ Class ____________

LESSON 4-3

Reading Strategies

Understand Labels

Examine these two triangles.

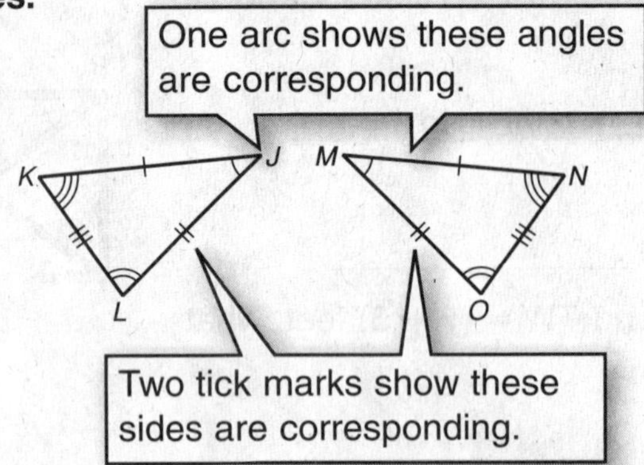

1. How can you tell which angle corresponds to $\angle L$?

____

2. How can you tell which side corresponds to $\overline{KL}$?

____

Answer the following questions based on these two triangles.

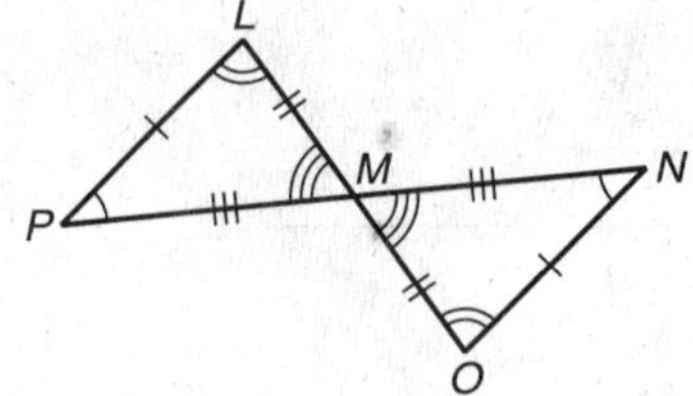

3. What angle corresponds to $\angle LMP$? ____________

4. What angle corresponds to $\angle P$? ____________

5. What side corresponds to $\overline{PL}$? ____________

6. What side corresponds to $\overline{LM}$? ____________

These two triangles are congruent. This statement can be written as follows: $\triangle ABC \cong \triangle XYZ$.
Labeling triangles in this way is meaningful because it states that in these two triangles, $\angle A \cong \angle X$; $\angle B \cong \angle Y$; and $\angle C \cong \angle Z$. The order in which the letters are placed tells which angles are congruent.

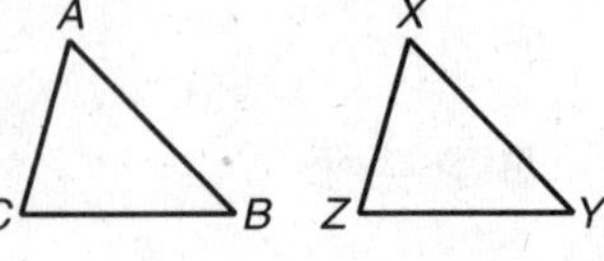

Answer the following questions based on these two triangles.

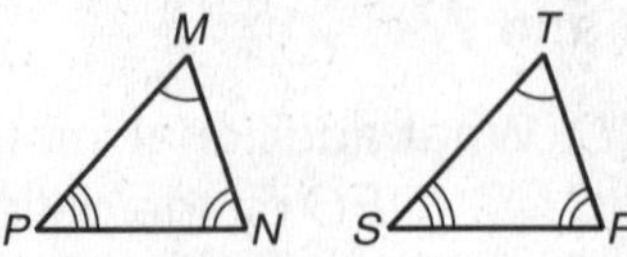

7. Write a congruence statement for these two triangles. ____________

8. How did you determine the order of the letters in your congruence statement?

____

____

Name ______________________ Date __________ Class __________

LESSON 4-4

Practice A

Triangle Congruence: SSS and SAS

Name the included angle for each pair of sides.

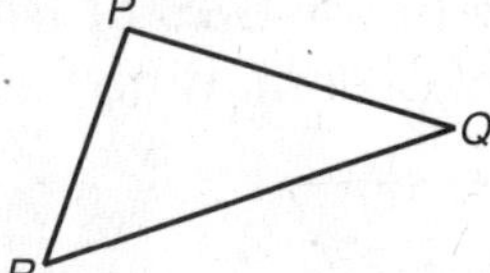

1. $\overline{PQ}$ and $\overline{PR}$ ________
2. $\overline{RQ}$ and $\overline{PR}$ ________
3. $\overline{PQ}$ and $\overline{RQ}$ ________

Write SSS (Side-Side-Side Congruence) or SAS (Side-Angle-Side Congruence) next to the correct postulate.

4. If three sides of one triangle are congruent to three sides of another triangle, then the triangles are congruent. ____________
5. If two sides and the included angle of one triangle are congruent to two sides and the included angle of another triangle, then the triangles are congruent. ____________

***GHIJ* is a rectangle. A rectangle is a four-sided figure with four right angles and congruent opposite sides. For Exercises 6 and 7, fill in the blanks to show that $\triangle GHJ \cong \triangle IJH$ in two different ways.**

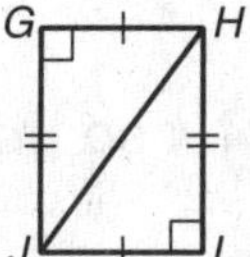

6. It is given that ________ and ________ are right angles and that $\overline{GH} \cong \overline{JI}$ and $\overline{GJ} \cong \overline{HI}$. All right angles are congruent. So $\triangle GHJ \cong \triangle IJH$ by ________.
7. It is given that $\overline{GH} \cong$ ________ and $\overline{GJ} \cong$ ________. By the Reflexive Property of $\cong$, $\overline{JH} \cong$ ________. So $\triangle GHJ \cong \triangle IJH$ by ________.
8. U.S. President Harry Truman and British Prime Minister Winston Churchill both wore polka-dot bow ties while in office. A well-tied bow tie resembles two congruent triangles. Use the phrases from the word bank to complete this two-column proof.

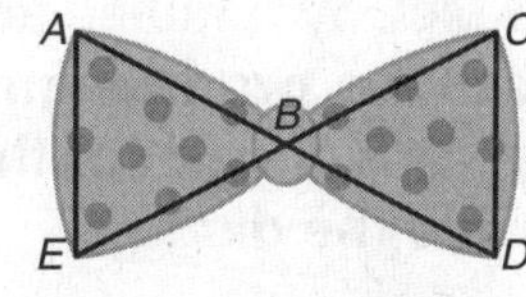

$\angle ABE \cong \angle DBC$,
SAS,
$\overline{BA} \cong \overline{BD}$,
$\overline{BE} \cong \overline{BC}$

Given: $\overline{BA} \cong \overline{BD}$, $\overline{BE} \cong \overline{BC}$

Prove: $\triangle ABE \cong \triangle DBC$

Proof:

Statements	Reasons
1. **a.** ______________	1. Given
2. **b.** ______________	2. Vert. ∠s Thm.
3. $\triangle ABE \cong \triangle DBC$	3. **c.** ______________

Name ________________________________ Date ____________ Class ____________

LESSON 4-4

Practice B

Triangle Congruence: SSS and SAS

Write which of the SSS or SAS postulates, if either, can be used to prove the triangles congruent. If no triangles can be proved congruent, write *neither*.

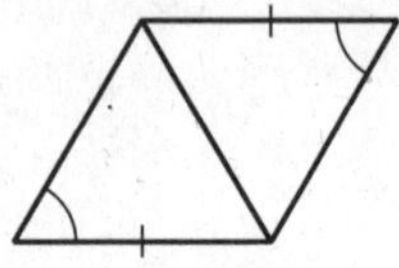

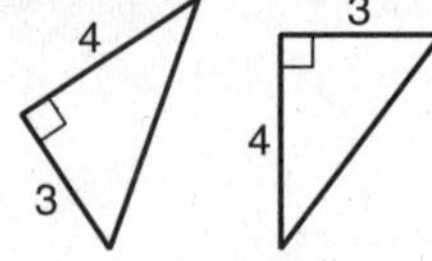

1. ____________________ **2.** ____________________

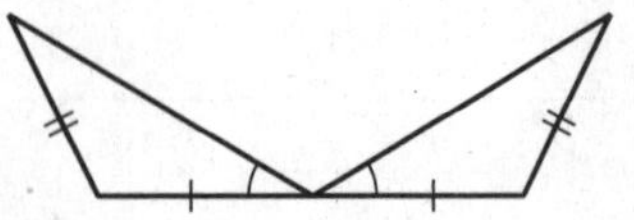

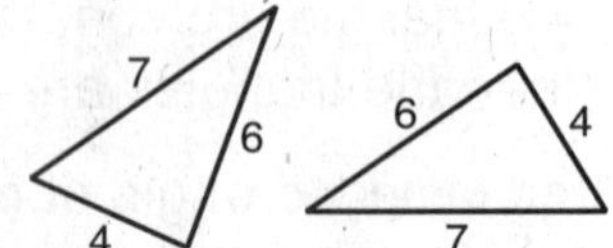

3. ____________________ **4.** ____________________

Find the value of x so that the triangles are congruent.

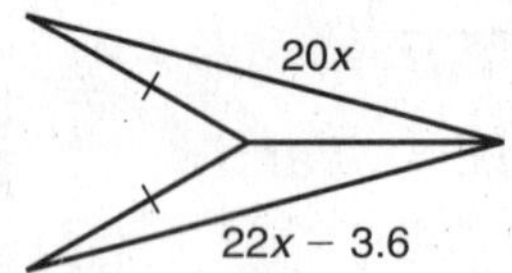

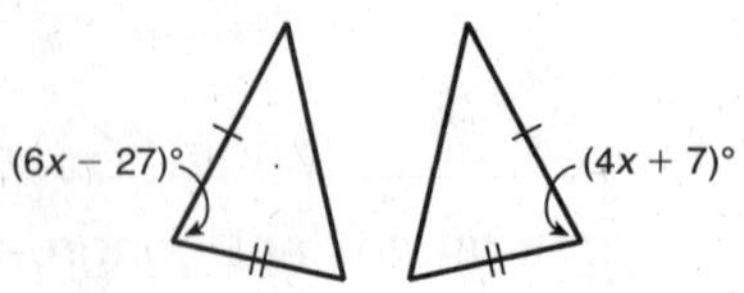

5. $x =$ ____________________ **6.** $x =$ ____________________

The Hatfield and McCoy families are feuding over some land. Neither family will be satisfied unless the two triangular fields are exactly the same size. You know that *C* is the midpoint of each of the intersecting segments. Write a two-column proof that will settle the dispute.

7. Given: C is the midpoint of $\overline{AD}$ and $\overline{BE}$.

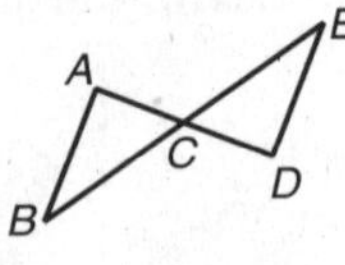

Prove: $\triangle ABC \cong \triangle DEC$

Proof:

Name ______________________ Date __________ Class __________

LESSON 4-4

Practice C

Triangle Congruence: SSS and SAS

For each definition, tell what angle measures or side lengths of the quadrilateral must be given in order to determine a specific figure. (*Hint:* Think in terms of congruent triangles.)

1. A *square* is a quadrilateral with four congruent sides and four right angles.

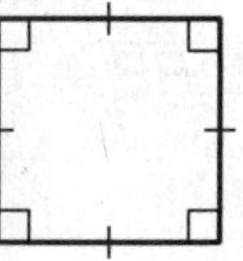

____

2. A *rectangle* is a quadrilateral with congruent pairs of opposite sides and four right angles.

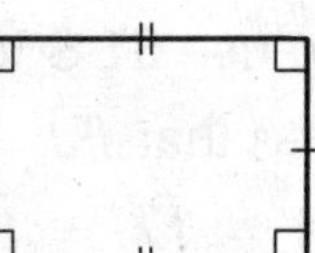

____

3. A *rhombus* is a quadrilateral with four congruent sides and parallel pairs of opposite sides.

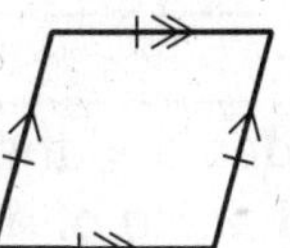

____

4. A *parallelogram* is a quadrilateral with congruent and parallel pairs of opposite sides.

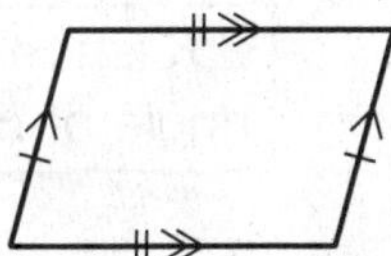

____

____

5. Can a square be determined given only the length of a diagonal? Explain your answer.

____

____

____

6. Find the total area of the figure.

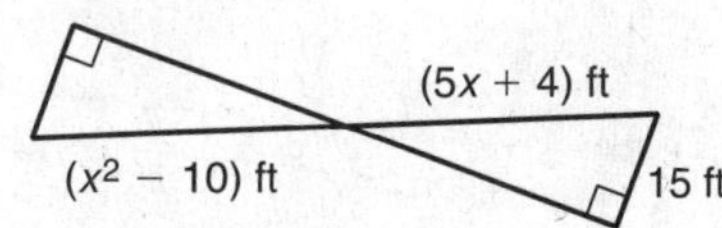

____

Write a paragraph proof.

7. **Given:** $\overline{BA} \cong \overline{BC}$, $\overline{BE} \cong \overline{BF}$, $\overline{AG} \cong \overline{CD}$, $\overline{GF} \cong \overline{DE}$

Prove: $\triangle AEG \cong \triangle CFD$

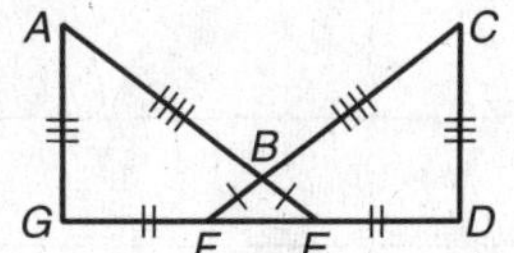

Name ______________________ Date ____________ Class ____________

LESSON 4-4

Reteach

Triangle Congruence: SSS and SAS

Side-Side-Side (SSS) Congruence Postulate

If three sides of one triangle are congruent to three sides of another triangle, then the triangles are congruent.

$\overline{QR} \cong \overline{TU}$, $\overline{RP} \cong \overline{US}$, and $\overline{PQ} \cong \overline{ST}$, so $\triangle PQR \cong \triangle STU$.

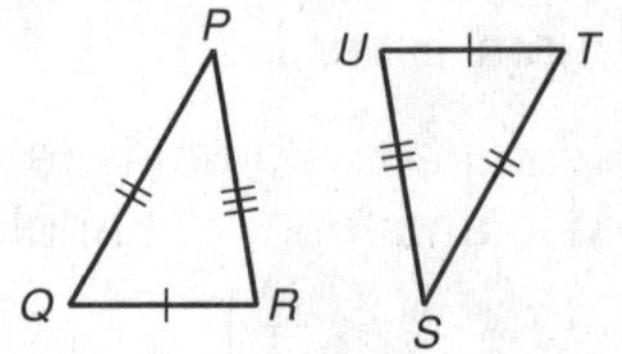

You can use SSS to explain why $\triangle FJH \cong \triangle FGH$.
It is given that $\overline{FJ} \cong \overline{FG}$ and that $\overline{JH} \cong \overline{GH}$. By the Reflex. Prop. of $\cong$, $\overline{FH} \cong \overline{FH}$. So $\triangle FJH \cong \triangle FGH$ by SSS.

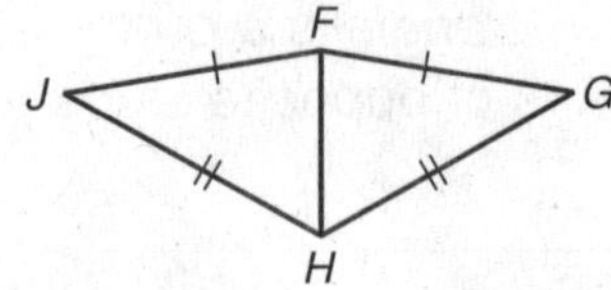

Side-Angle-Side (SAS) Congruence Postulate

If two sides and the included angle of one triangle are congruent to two sides and the included angle of another triangle, then the triangles are congruent.

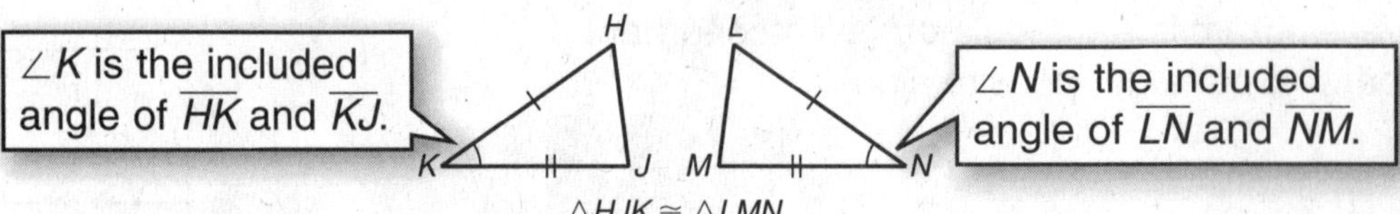

Use SSS to explain why the triangles in each pair are congruent.

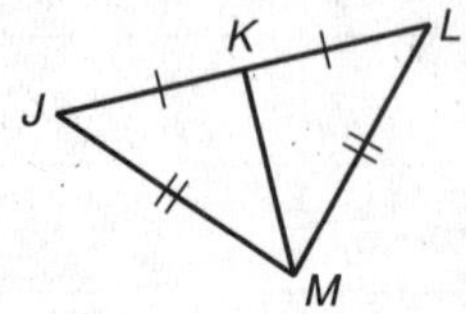

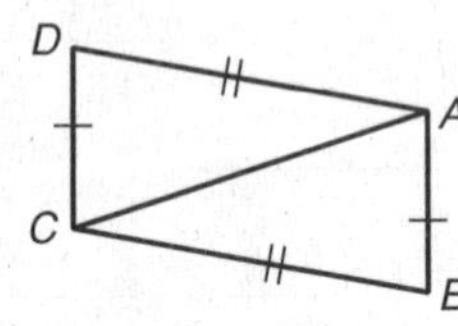

1. $\triangle JKM \cong \triangle LKM$

____

____

____

____

2. $\triangle ABC \cong \triangle CDA$

____

____

____

____

3. Use SAS to explain why $\triangle WXY \cong \triangle WZY$.

____

____

____

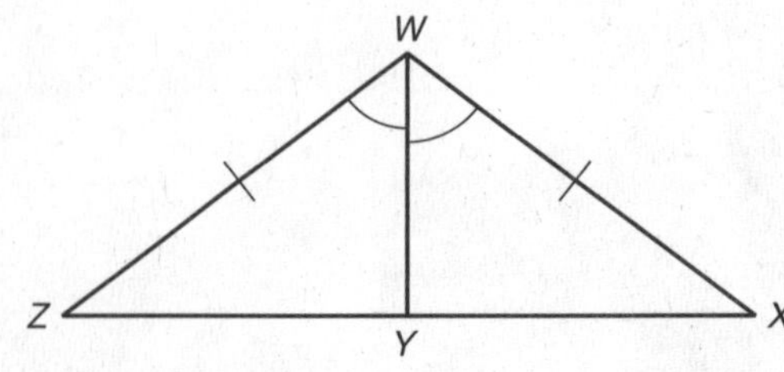

Name ______________________ Date __________ Class __________

LESSON 4-4

Reteach

Triangle Congruence: SSS and SAS continued

You can show that two triangles are congruent by using SSS and SAS.

Show that $\triangle JKL \cong \triangle FGH$ for $y = 7$.

$HG = y + 6$ $\quad m\angle G = 5y + 5$ $\quad FG = 4y - 1$

$= 7 + 6 = 13$ $\quad = 5(7) + 5 = 40°$ $\quad = 4(7) - 1 = 27$

$HG = LK = 13$, so $\overline{HG} \cong \overline{LK}$ by def. of $\cong$ segs. $m\angle G = 40°$, so $\angle G \cong \angle K$ by def. of $\cong$ $\angle$s. $FG = JK = 27$, so $\overline{FG} \cong \overline{JK}$ by def. of $\cong$ segs. Therefore $\triangle JKL \cong \triangle FGH$ by SAS.

Show that the triangles are congruent for the given value of the variable.

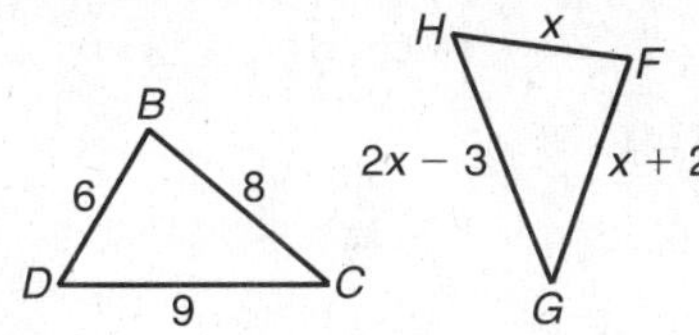

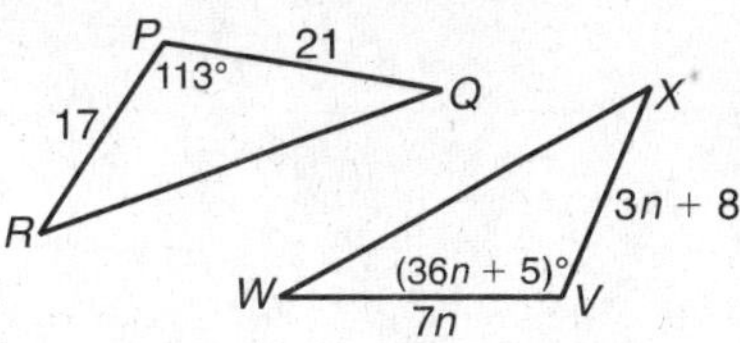

4. $\triangle BCD \cong \triangle FGH$, $x = 6$

____

____

____

____

____

____

5. $\triangle PQR \cong \triangle VWX$, $n = 3$

____

____

____

____

____

____

6. Complete the proof.

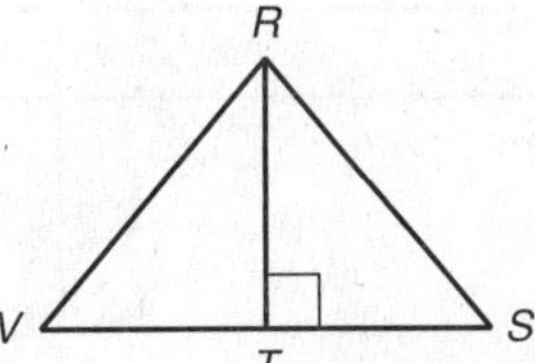

Given: T is the midpoint of $\overline{VS}$.

$\overline{RT} \perp \overline{VS}$

Prove: $\triangle RST \cong \triangle RVT$

Statements	Reasons
1. T is the midpoint of $\overline{VS}$.	1. Given
2. **a.** ______________	2. Def. of mdpt.
3. $\overline{RT} \perp \overline{VS}$	3. **b.** ______________
4. ______________	4. **c.** ______________
5. **d.** ______________	5. Rt. $\angle \cong$ Thm.
6. $\overline{RT} \cong \overline{RT}$	6. **e.** ______________
7. $\triangle RST \cong \triangle RVT$	7. **f.** ______________

Name ______________________ Date __________ Class __________

LESSON 4-4

Challenge

More Proofs with Overlapping Triangles

Write each two-column proof.

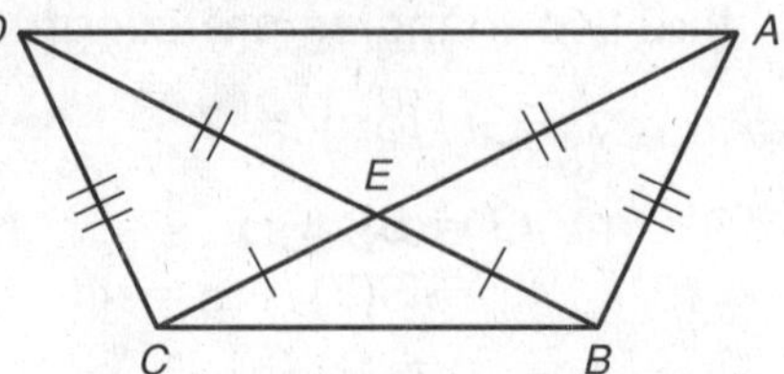

1. Given: $\overline{CE} \cong \overline{BE}$, $\overline{EA} \cong \overline{ED}$, $\overline{DC} \cong \overline{AB}$

Prove: $\triangle ABC \cong \triangle DCB$

Proof:

Statements	Reasons

2. Given: $\overline{LK} \cong \overline{HJ}$, $\overline{GK} \cong \overline{GJ}$, $\angle GKL \cong \angle GJH$

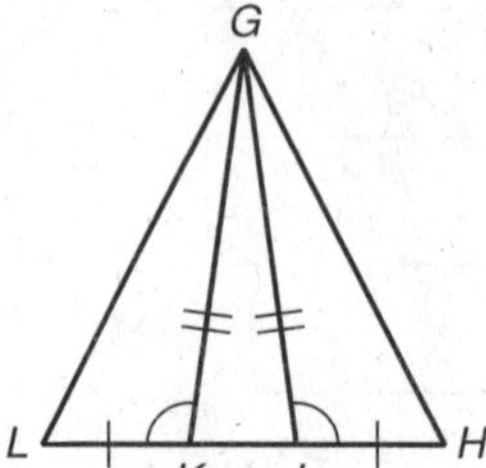

Prove: $\triangle GLJ \cong \triangle GHK$

Proof:

Statements	Reasons

Name ________________________ Date ____________ Class ____________

LESSON 4-4

Problem Solving

Triangle Congruence: SSS and SAS

Use the diagram for Exercises 1 and 2.

A shed door appears to be divided into congruent right triangles.

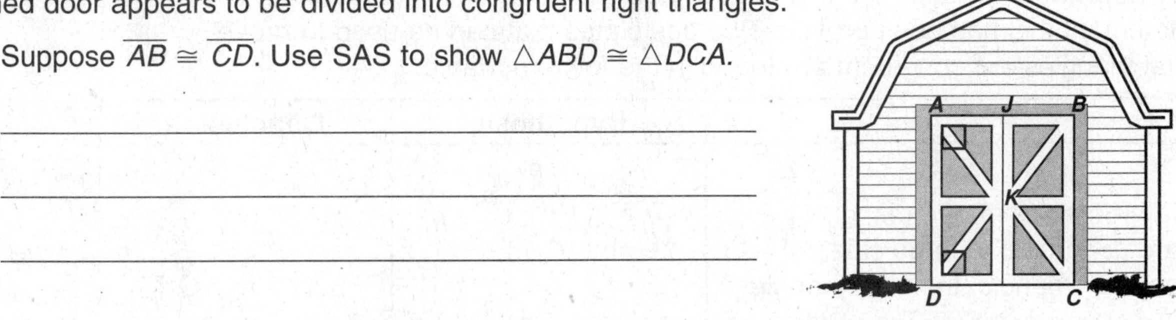

1. Suppose $\overline{AB} \cong \overline{CD}$. Use SAS to show $\triangle ABD \cong \triangle DCA$.

____

____

____

____

2. *J* is the midpoint of *AB* and $\overline{AK} \cong \overline{BK}$. Use SSS to explain why $\triangle AKJ \cong \triangle BKJ$.

____

____

____

3. A *balalaika* is a Russian stringed instrument. Show that the triangular parts of the two balalaikas are congruent for $x = 6$.

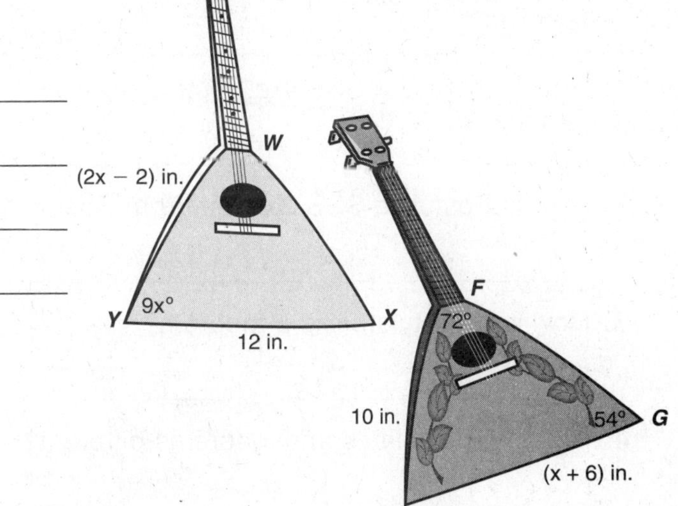

____

____

____

____

A quilt pattern of a dog is shown. Choose the best answer.

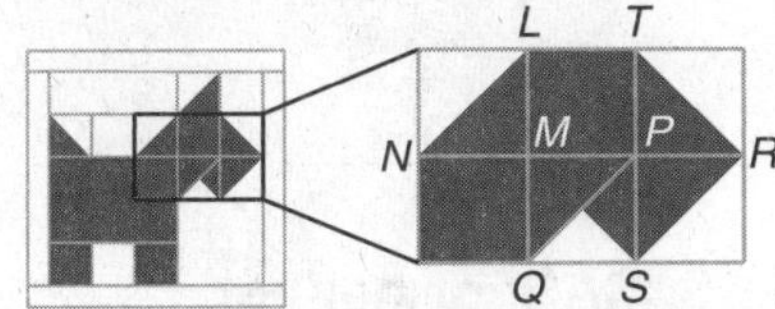

4. $ML = MP = MN = MQ = 1$ inch. Which statement is correct?

A $\triangle LMN \cong \triangle QMP$ by SAS.

B $\triangle LMN \cong \triangle QMP$ by SSS.

C $\triangle LMN \cong \triangle MQP$ by SAS.

D $\triangle LMN \cong \triangle MQP$ by SSS.

5. *P* is the midpoint of $\overline{TS}$ and $TR = SR = 1.4$ inches. What can you conclude about $\triangle TRP$ and $\triangle SRP$?

F $\triangle TRP \cong \triangle SRP$ by SAS.

G $\triangle TRP \cong \triangle SRP$ by SSS.

H $\triangle TRP \cong \triangle SPR$ by SAS.

J $\triangle TRP \cong \triangle SPR$ by SSS.

Name ______________________________ Date ____________ Class ____________

LESSON 4-4

Reading Strategies
Compare and Contrast

In mathematics, postulates and theorems are used to explain relationships. A **postulate** is a statement that is accepted without proof. A **theorem** is a statement that has been proven. Two postulates that can be used to prove that triangles are congruent are found in the following table:

Postulate	Hypothesis	Conclusion
SSS If three sides of one triangle are congruent to three sides of another triangle, then the triangles are congruent: Side-Side-Side Congruence	$\overline{XY} \cong \overline{QR}$ $\overline{XZ} \cong \overline{QS}$ $\overline{YZ} \cong \overline{RS}$	$\triangle XYZ \cong \triangle QRS$
SAS If two sides and the included angle of one triangle are congruent to two sides and the included angle of another triangle, then the triangles are congruent: Side-Angle-Side Congruence	$\overline{LM} \cong \overline{TU}$ $\overline{LN} \cong \overline{TV}$ $\angle NLM \cong \angle VTU$	$\triangle LMN \cong \triangle TUV$

1. How is Postulate SSS like Postulate SAS?

____

2. How is Postulate SSS different from Postulate SAS?

____

3. How is a postulate like a theorem?

____

4. How are postulates and theorems different?

____

Determine whether each pair of triangles is congruent by SSS, SAS, or neither.

5. 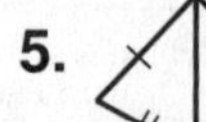 

____

6. 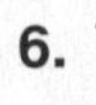 

____

7.

____

8.

____

Name ______________________ Date __________ Class __________

LESSON 4-5

Practice A

Triangle Congruence: ASA, AAS, and HL

Name the included side for each pair of consecutive angles.

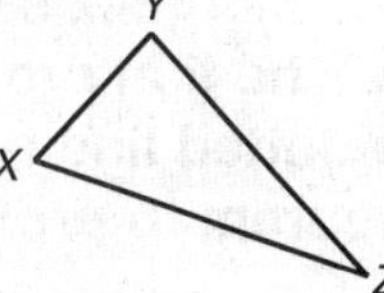

1. $\angle X$ and $\angle Z$ ________ **2.** $\angle Y$ and $\angle X$ ________

3. $\angle Y$ and $\angle Z$ ________

Write *ASA* (Angle-Side-Angle Congruence), *AAS* (Angle-Angle-Side Congruence), or *HL* (Hypotenuse-Leg Congruence) next to the correct postulate.

4. If the hypotenuse and a leg of one right triangle are congruent to the hypotenuse and a leg of another right triangle, then the triangles are congruent. ________

5. If two angles and a nonincluded side of one triangle are congruent to the corresponding angles and nonincluded side of another triangle, then the triangles are congruent. ________

6. If two angles and the included side of one triangle are congruent to two angles and the included side of another triangle, then the triangles are congruent. ________

For Exercises 7–9, tell whether you can use each congruence theorem to prove that $\triangle ABC \cong \triangle DEF$. If not, tell what else you need to know.

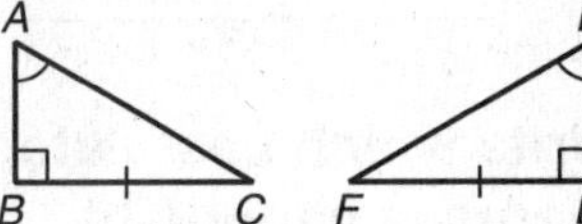

7. Hypotenuse-Leg

____

8. Angle-Side-Angle

____

9. Angle-Angle-Side

____

10. A standard letter-sized envelope is a $9\frac{1}{2}$-in.-by-4-in. rectangle. The envelope is folded and glued from a sheet of paper shaped like the figure. Use the phrases in the word bank to complete this proof.

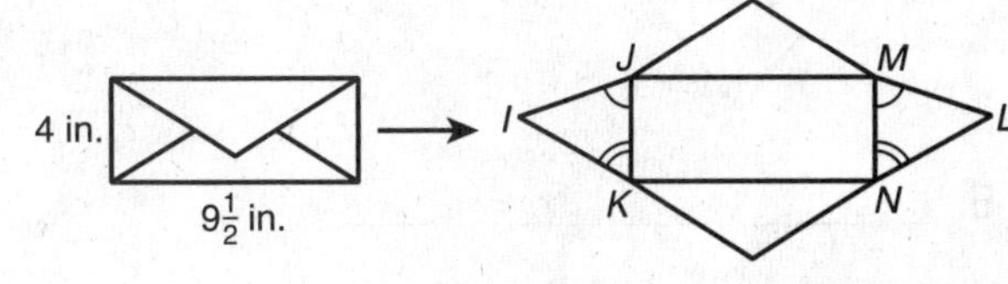

Given,
ASA,
Definition of rectangle

Given: *JMNK* is a rectangle. $\angle IJK \cong \angle LMN$, $\angle IKJ \cong \angle LNM$

Prove: $\triangle IJK \cong \triangle LMN$

Statements	Reasons
1. $\angle IJK \cong \angle LMN$, $\angle IKJ \cong \angle LNM$	1. **a.** ____________
2. $\overline{JK} \cong \overline{MN}$	2. **b.** ____________
3. $\triangle IJK \cong \triangle LMN$	3. **c.** ____________

Name ______________________________ Date ____________ Class ____________

LESSON **4-5**

Practice B

Triangle Congruence: ASA, AAS, and HL

Students in Mrs. Marquez's class are watching a film on the uses of geometry in architecture. The film projector casts the image on a flat screen as shown in the figure. The dotted line is the bisector of $\angle ABC$. Tell whether you can use each congruence theorem to prove that $\triangle ABD \cong \triangle CBD$. If not, tell what else you need to know.

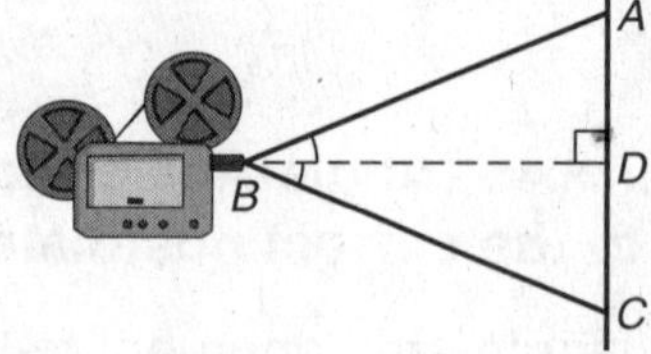

1. Hypotenuse-Leg

____

2. Angle-Side-Angle

____

3. Angle-Angle-Side

____

Write which postulate, if any, can be used to prove the pair of triangles congruent.

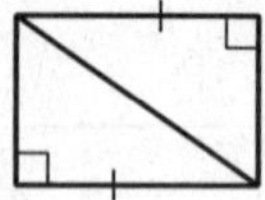

4. ____________________

5. ____________________

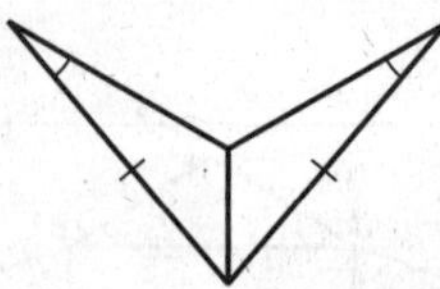

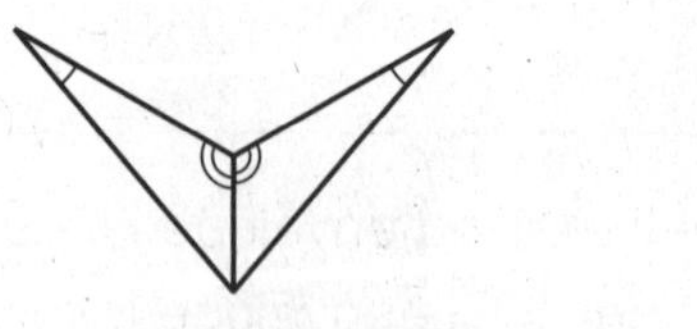

6. ____________________

7. ____________________

Write a paragraph proof.

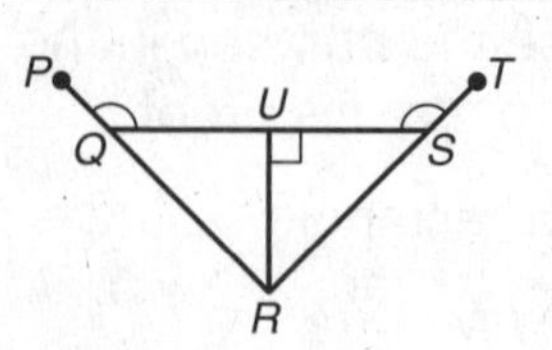

8. **Given:** $\angle PQU \cong \angle TSU$,
$\angle QUR$ and $\angle SUR$ are right angles.

Prove: $\triangle RUQ \cong \triangle RUS$

Name ______________________ Date __________ Class __________

LESSON 4-5

Practice C

Triangle Congruence: ASA, AAS, and HL

1. In what case is knowing three parts (side lengths or angle measures) of a triangle not sufficient information to determine a specific triangle?

____

____

2. Describe how the Leg-Leg (LL), Hypotenuse-Angle (HA), and Leg-Angle (LA) theorems of right-triangle congruence are related to the theorems and postulates of congruence that apply to all triangles.

____

____

____

____

____

3. You know that a triangle is a right triangle. You are given two more parts. Is it possible that you do not have sufficient information to determine a specific triangle? Explain your answer.

____

____

Write a paragraph proof. Remember that if triangles are congruent, their matching parts are congruent.

4. Given: $\overline{GB}$, $\overline{GD}$, and $\overline{GF}$ are radii of the circle centered at G and are perpendicular to the sides of $\triangle ACE$.

Prove: $\triangle ACE$ is equilateral.

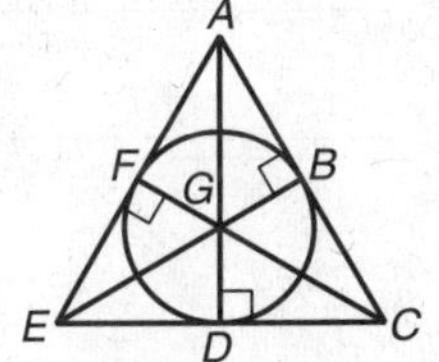

Name ______________________ Date ____________ Class ____________

LESSON 4-5

Reteach

Triangle Congruence: ASA, AAS, and HL

Angle-Side-Angle (ASA) Congruence Postulate

If two angles and the included side of one triangle are congruent to two angles and the included side of another triangle, then the triangles are congruent.

$\overline{AC}$ is the included side of $\angle A$ and $\angle C$.

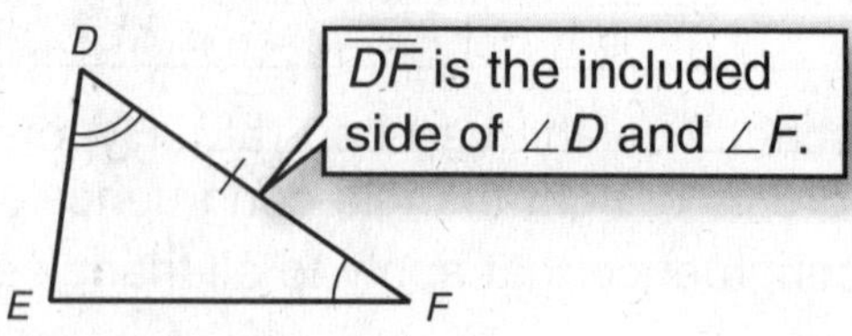

$\triangle ABC \cong \triangle DEF$

Determine whether you can use ASA to prove the triangles congruent. Explain.

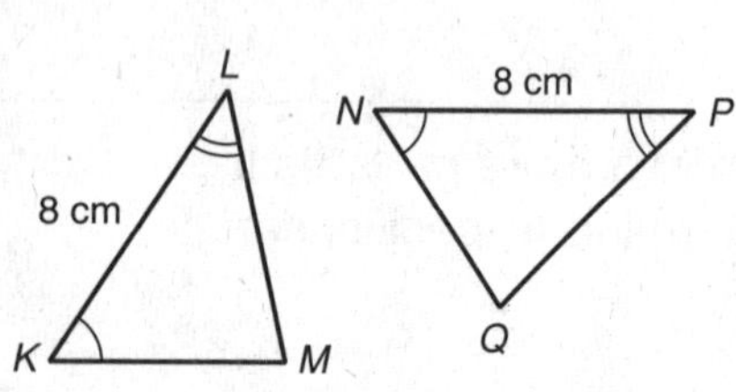

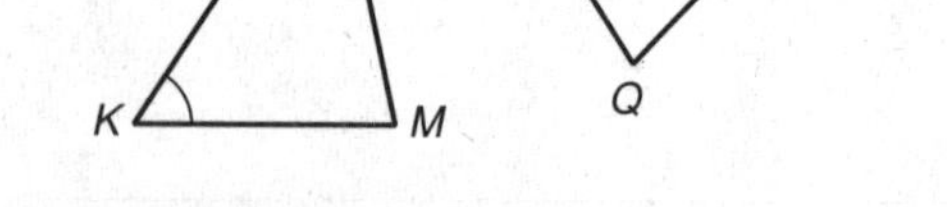

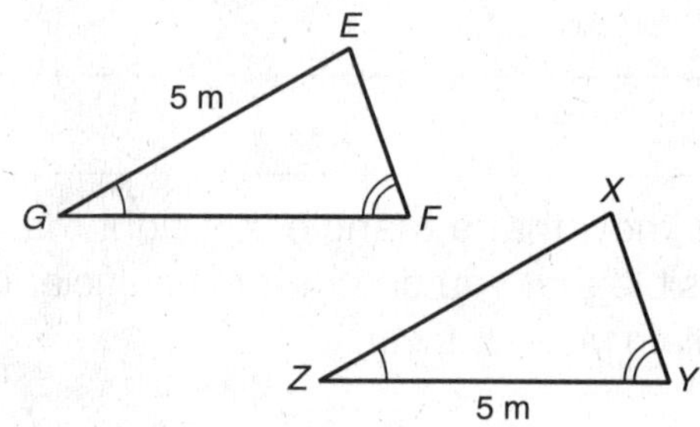

1. $\triangle KLM$ and $\triangle NPQ$

2. $\triangle EFG$ and $\triangle XYZ$

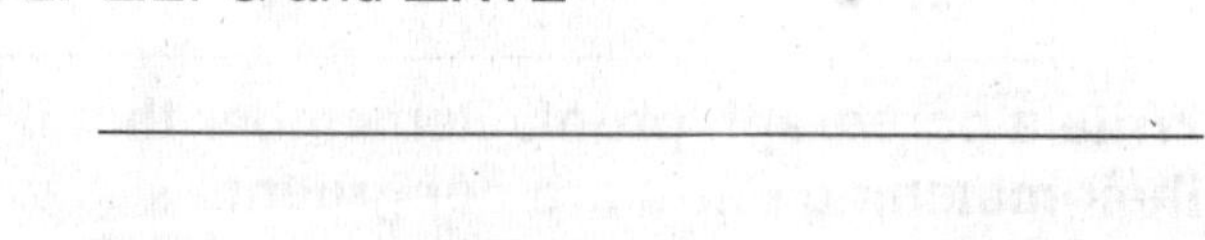

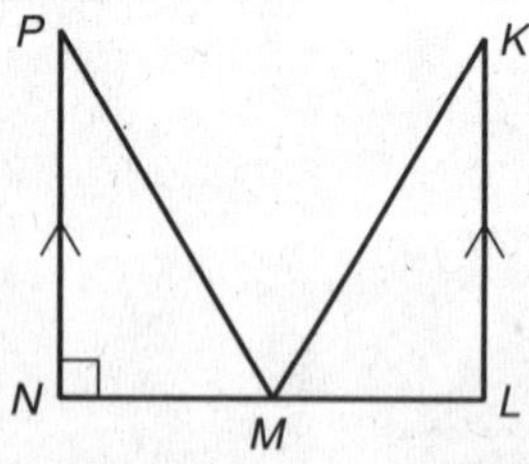

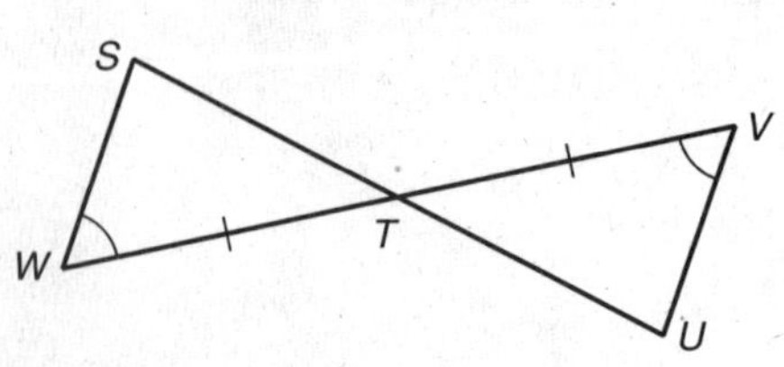

3. $\triangle KLM$ and $\triangle PNM$, given that M is the midpoint of $\overline{NL}$

4. $\triangle STW$ and $\triangle UTV$

Name ______________________ Date __________ Class __________

LESSON 4-5

Reteach

Triangle Congruence: ASA, AAS, and HL continued

Angle-Angle-Side (AAS) Congruence Theorem

If two angles and a nonincluded side of one triangle are congruent to the corresponding angles and nonincluded side of another triangle, then the triangles are congruent.

$\overline{FH}$ is a nonincluded side of $\angle F$ and $\angle G$.

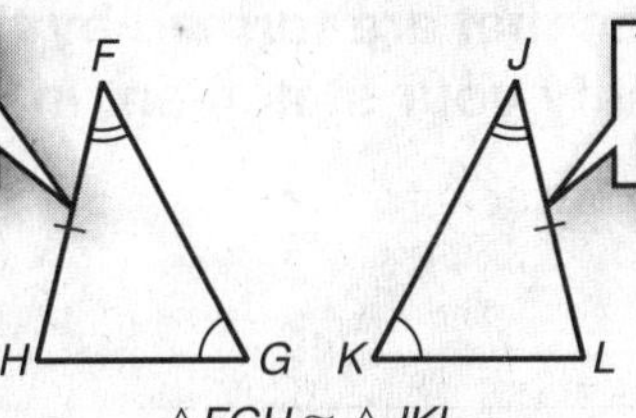

$\overline{JL}$ is a nonincluded side of $\angle J$ and $\angle K$.

$\triangle FGH \cong \triangle JKL$

Special theorems can be used to prove right triangles congruent.

Hypotenuse-Leg (HL) Congruence Theorem

If the hypotenuse and a leg of a right triangle are congruent to the hypotenuse and a leg of another right triangle, then the triangles are congruent.

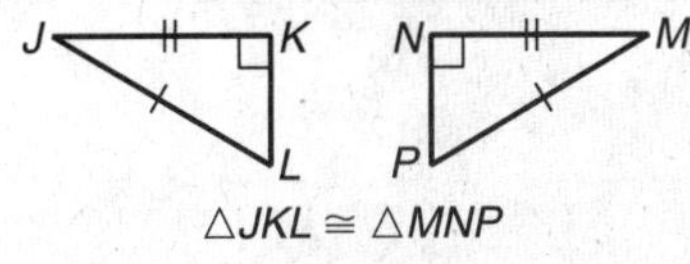

$\triangle JKL \cong \triangle MNP$

5. Describe the corresponding parts and the justifications for using them to prove the triangles congruent by AAS.

Given: $\overline{BD}$ is the angle bisector of $\angle ADC$.

Prove: $\triangle ABD \cong \triangle CBD$

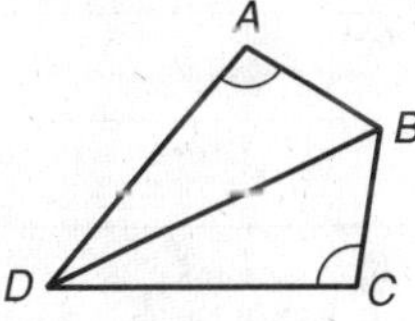

____

____

Determine whether you can use the HL Congruence Theorem to prove the triangles congruent. If yes, explain. If not, tell what else you need to know.

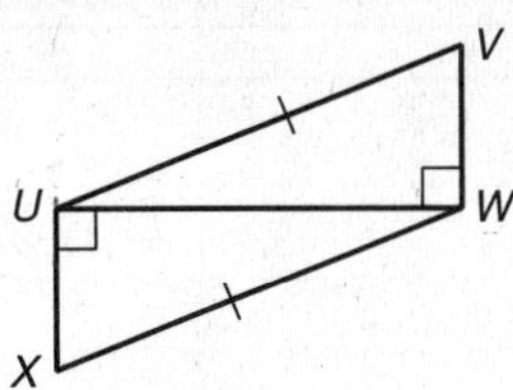

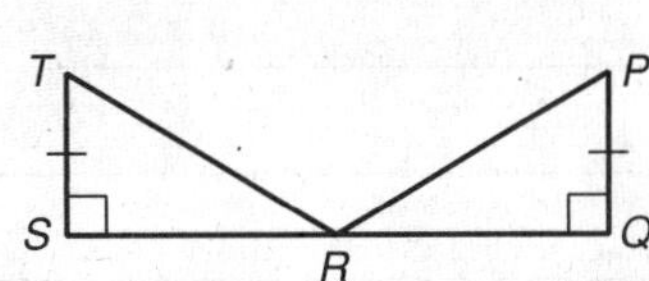

6. $\triangle UVW \cong \triangle WXU$

____

____

7. $\triangle TSR \cong \triangle PQR$

____

____

Name ______________________ Date __________ Class __________

LESSON 4-5

Challenge

Congruent Triangles on the Coordinate Plane

When proving that two triangles are congruent, you may need to use your knowledge of lines on the coordinate plane.

Graph each set of lines. Then:
a. Identify two congruent triangles that are formed by the lines.
b. Write a paragraph proof to justify your statement in part a.

1. $y = 5$ $\quad y = -x + 1$
$y = -4$ $\quad y = -3x + 2$

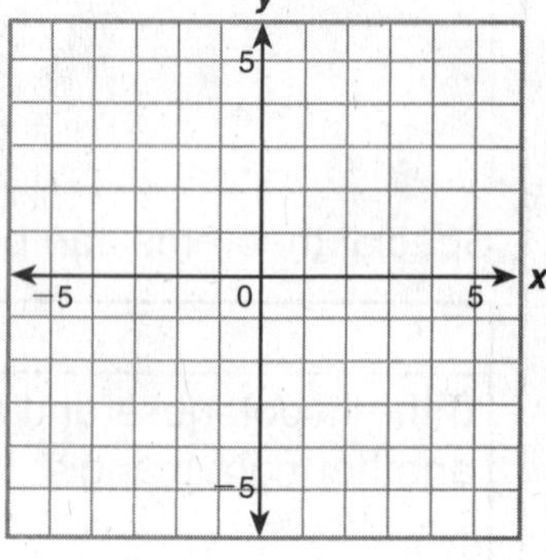

a. ______________________

b. ______________________

2. $y = 3$ $\quad y = 2x + 1$
$x = -1$ $\quad y = -2x - 3$

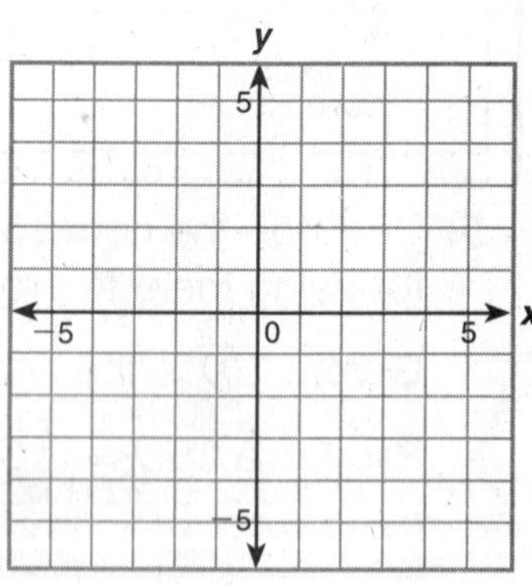

a. ______________________

b. ______________________

3. $y = -4$ $\quad y = x - 2$
$x = -2$ $\quad y = -x$

a. ______________________

b. ______________________

4. On a separate sheet of paper, write equations for a different set of lines that meet to form congruent triangles. Identify the congruent triangles, and write a proof to justify the congruence.

Name ______________________________ Date ____________ Class ____________

LESSON 4-5

Problem Solving

Triangle Congruence: ASA, AAS, and HL

Use the following information for Exercises 1 and 2.

Melanie is at hole 6 on a miniature golf course. She walks east 7.5 meters to hole 7. She then faces south, turns 67° west, and walks to hole 8. From hole 8, she faces north, turns 35° west, and walks to hole 6.

1. Draw the section of the golf course described. Label the measures of the angles in the triangle.

2. Is there enough information given to determine the location of holes 6, 7, and 8? Explain.

____

3. A section of the front of an English Tudor home is shown in the diagram. If you know that $\overline{KN} \cong \overline{LN}$ and $\overline{JN} \cong \overline{MN}$, can you use HL to conclude that $\triangle JKN \cong \triangle MLN$? Explain.

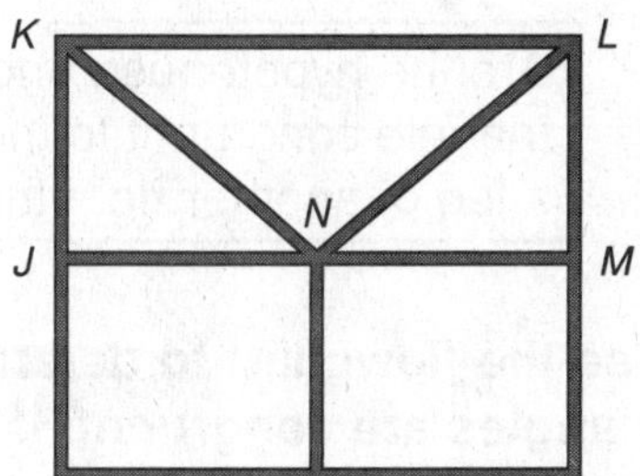

____

____

Use the diagram of a kite for Exercises 4 and 5.

$\overline{AE}$ is the angle bisector of $\angle DAF$ and $\angle DEF$.

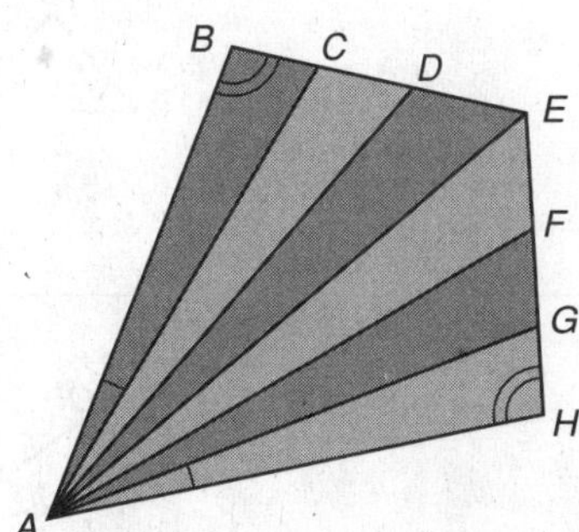

4. What can you conclude about $\triangle DEA$ and $\triangle FEA$?

   A $\triangle DEA \cong \triangle FEA$ by HL.

   B $\triangle DEA \cong \triangle FEA$ by AAA.

   C $\triangle DEA \cong \triangle FEA$ by ASA.

   D $\triangle DEA \cong \triangle FEA$ by SAS.

5. Based on the diagram, what can you conclude about $\triangle BCA$ and $\triangle HGA$?

   F $\triangle BCA \cong \triangle HGA$ by HL.

   G $\triangle BCA \cong \triangle HGA$ by AAS.

   H $\triangle BCA \cong \triangle HGA$ by ASA.

   J It cannot be shown using the given information that $\triangle BCA \cong \triangle HGA$.

Name ______________________ Date __________ Class __________

LESSON 4-5

Reading Strategies

Use a Graphic Aid

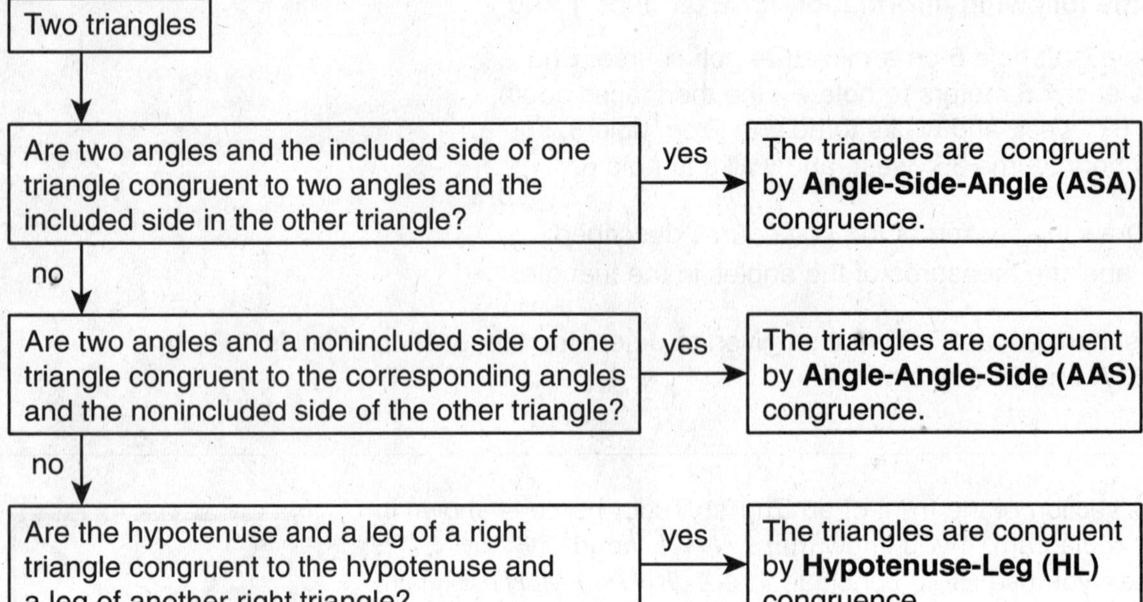

Use the flowchart to determine, if possible, whether the following pairs of triangles are congruent. If congruent, write *ASA*, *AAS*, or *HL*—the postulate you used to conclude that they are congruent. If it is not possible to conclude that they are congruent, write *no conclusion*.

1. 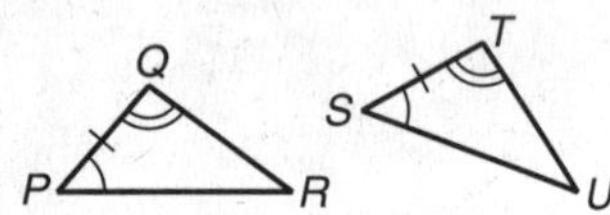

____

2. 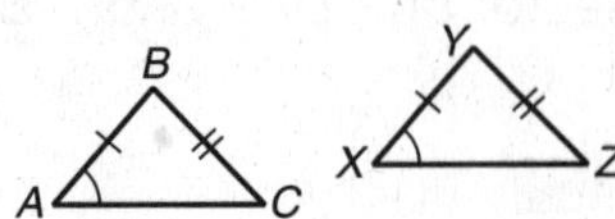

____

3. 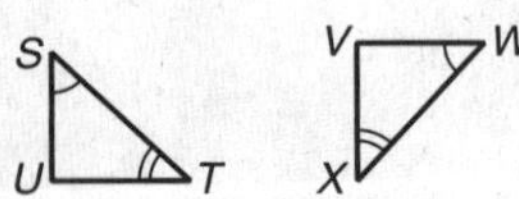

____

4. 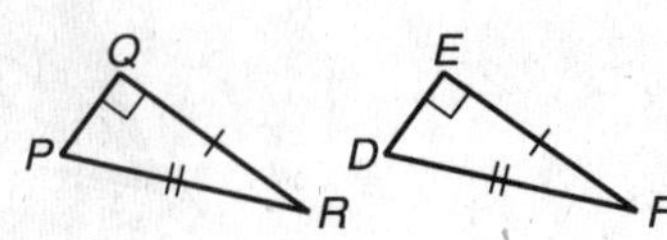

____

5. 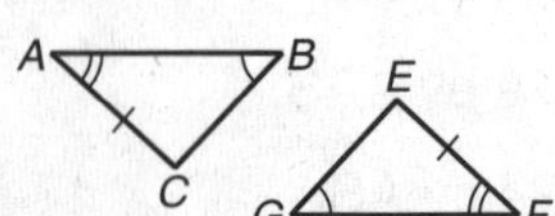

____

6. 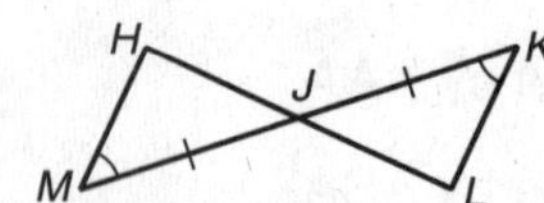

____

Name ______________________________ Date ____________ Class ____________

LESSON 4-6

Practice A

Triangle Congruence: CPCTC

1. CPCTC is an abbreviation of the phrase "Corresponding ____________ of Congruent ____________ are Congruent."

Use the figure for Exercises 2 and 3.
∠B ≅ ∠E, ∠C ≅ ∠F, and $\overline{AB} \cong \overline{DE}$.

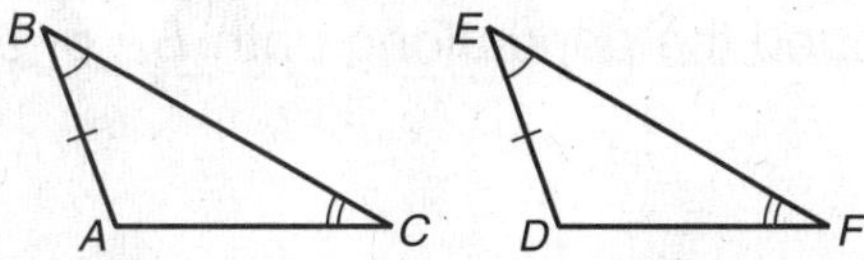

2. Name the triangle congruence theorem that shows △ABC ≅ △DEF. ____________

3. Use CPCTC to name the other three pairs of congruent parts in the triangles.

_______ ≅ _______ _______ ≅ _______ _______ ≅ _______

4. Some hikers come to a river in the woods. They want to cross the river but decide to find out how wide it is first. So they set up congruent right triangles. The figure shows the river and the triangles. Find the width of the river, GH. ____________

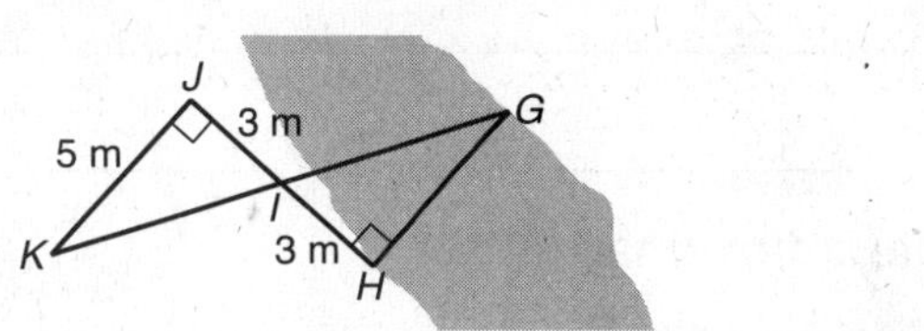

Use the phrases in the word bank to complete this proof.

5. **Given:** $\overline{PQ} \cong \overline{RQ}$, ∠PQS ≅ ∠RQS

Prove: ∠P ≅ ∠R

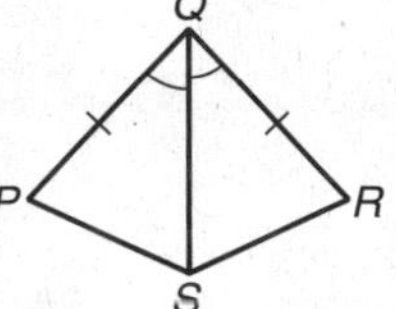

$\overline{QS} \cong \overline{QS}$,
CPCTC,
△PQS ≅ △RQS

Statements	Reasons
1. $\overline{PQ}$ ≅ RQ, ∠PQS ≅ ∠RQS	1. Given
2. **a.** ____________	2. Reflexive Property of ≅
3. **b.** ____________	3. SAS
4. ∠P ≅ ∠R	4. **c.** ____________

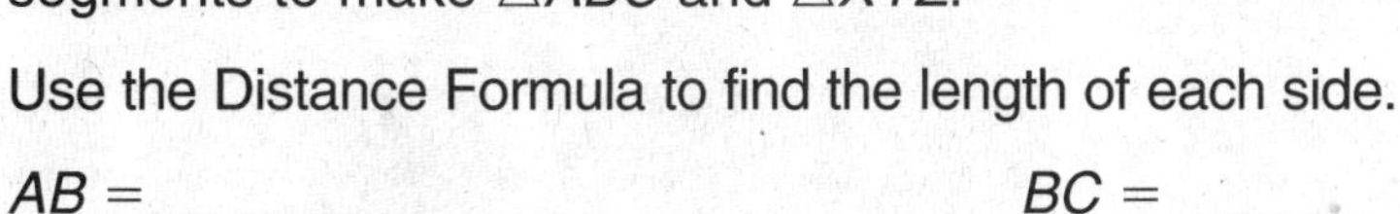

Use the blank graph for Exercises 6–9.

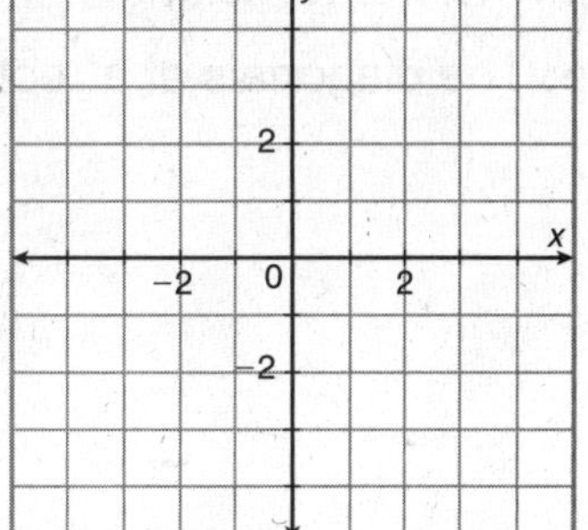

6. Plot these points: A(0, 0), B(0, 3), C(4, 0), X(−4, −3), Y(−4, 0), Z(0, −3). Draw segments to make △ABC and △XYZ.

7. Use the Distance Formula to find the length of each side.

AB = ________ BC = ________ AC = ________

XY = ________ YZ = ________ XZ = ________

8. Name the triangle congruence theorem that shows △ABC ≅ △XYZ. ________

9. How do you know that ∠X ≅ ∠A? ________

Name ________________________ Date ____________ Class ____________

LESSON 4-6

Practice B

Triangle Congruence: CPCTC

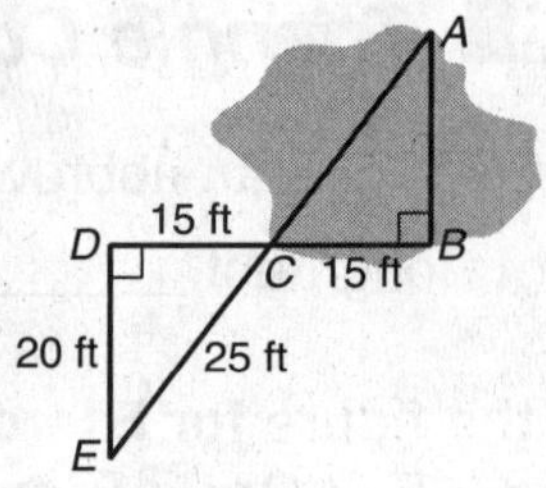

1. Heike Dreschler set the Woman's World Junior Record for the long jump in 1983. She jumped about 23.4 feet. The diagram shows two triangles and a pond. Explain whether Heike could have jumped the pond along path *BA* or along path *CA*.

____

____

____

____

____

Write a flowchart proof.

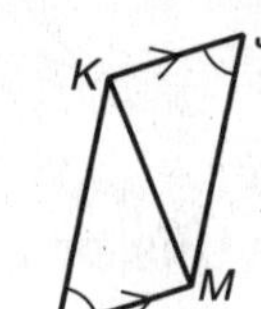

2. **Given:** $\angle L \cong \angle J$, $\overline{KJ} \parallel \overline{LM}$

   Prove: $\angle LKM \cong \angle JMK$

Write a two-column proof.

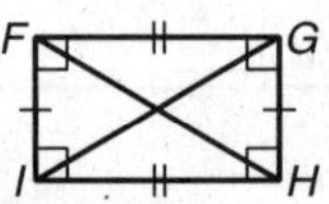

3. **Given:** *FGHI* is a rectangle.

   Prove: The diagonals of a rectangle have equal lengths.

Name ______________________________ Date ____________ Class ____________

LESSON 4-6

Practice C

Triangle Congruence: CPCTC

Write paragraph proofs for Exercises 1 and 2.

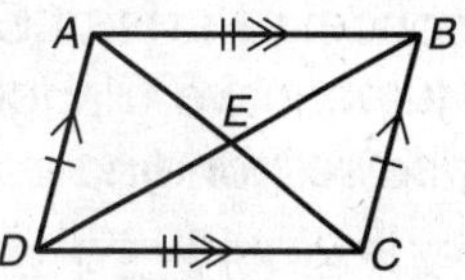

1. Given: *ABCD* is a parallelogram.

Prove: The diagonals of a parallelogram bisect each other.

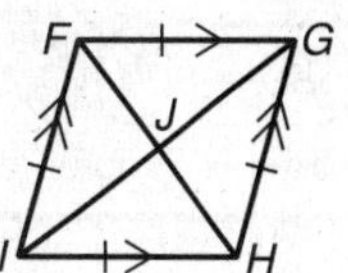

2. Given: *FGHI* is a rhombus.

Prove: The diagonals of a rhombus are congruent, perpendicular, and bisect the vertex angles of the rhombus.

(*Note:* Be careful naming the triangles. The order of vertices matters.)

3. Rectangles, rhombuses, and squares are all types of parallelograms. Write a conjecture about the diagonals of a rectangle.

____

4. A square is a type of rhombus. Write a conjecture about the diagonals of a square.

____

____

5. An isosceles trapezoid has one pair of noncongruent parallel sides, a pair of congruent nonparallel sides, and two pairs of congruent angles. What relationship do the diagonals of an isosceles trapezoid have?

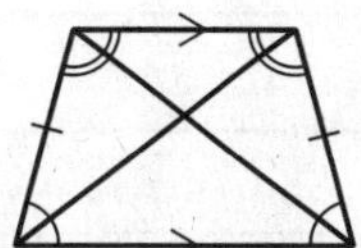

____

Name ______________________________ Date ____________ Class ____________

LESSON 4-6

Reteach

Triangle Congruence: CPCTC

Corresponding Parts of Congruent Triangles are Congruent **(CPCTC)** is useful in proofs. If you prove that two triangles are congruent, then you can use CPCTC as a justification for proving corresponding parts congruent.

Given: $\overline{AD} \cong \overline{CD}$, $\overline{AB} \cong \overline{CB}$

Prove: $\angle A \cong \angle C$

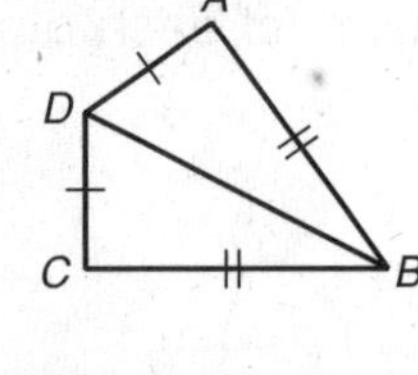

Proof:

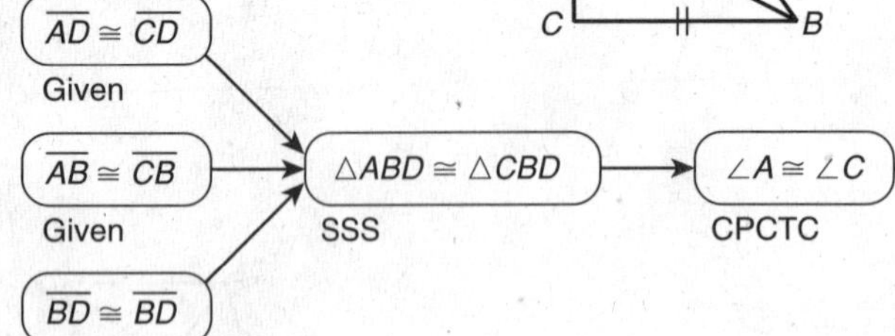

Complete each proof.

1. Given: $\angle PNQ \cong \angle LNM$, $\overline{PN} \cong \overline{LN}$, N is the midpoint of $\overline{QM}$.

Prove: $\overline{PQ} \cong \overline{LM}$

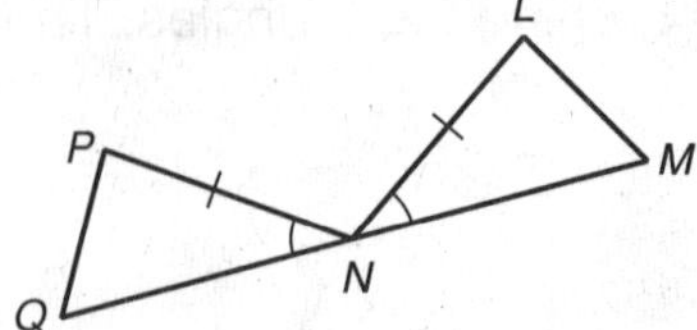

Proof:

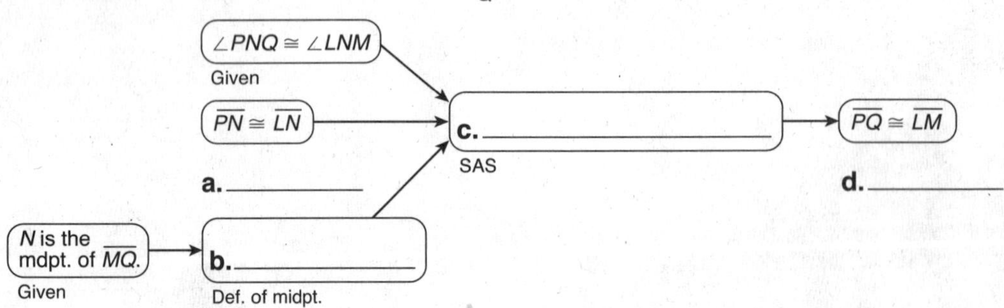

2. Given: $\triangle UXW$ and $\triangle UVW$ are right $\triangle$s.
$\overline{UX} \cong \overline{UV}$

Prove: $\angle X \cong \angle V$

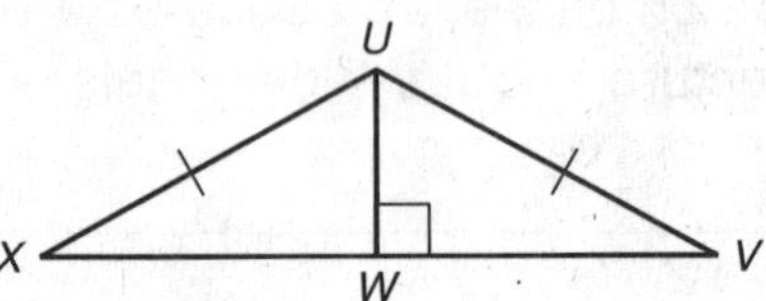

Proof:

Statements	**Reasons**
1. $\triangle UXW$ and $\triangle UVW$ are rt. $\triangle$s.	1. Given
2. $\overline{UX} \cong \overline{UV}$	2. **a.** ________________
3. $\overline{UW} \cong \overline{UW}$	3. **b.** ________________
4. **c.** ________________	4. **d.** ________________
5. $\angle X \cong \angle V$	5. **e.** ________________

Name ______________________ Date __________ Class __________

LESSON 4-6

Reteach

Triangle Congruence: CPCTC continued

You can also use CPCTC when triangles are on the coordinate plane.

Given: $C(2, 2)$, $D(4, -2)$, $E(0, -2)$, $F(0, 1)$, $G(-4, -1)$, $H(-4, 3)$

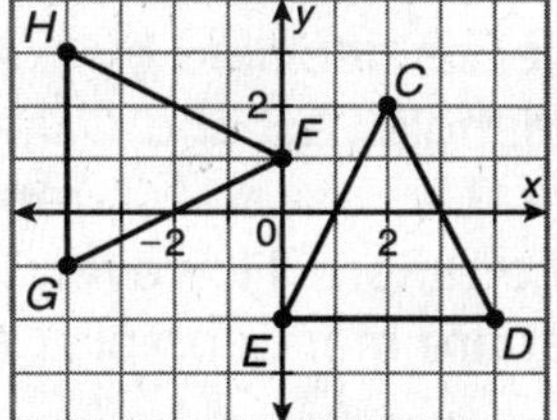

Prove: $\angle CED \cong \angle FHG$

Step 1 Plot the points on a coordinate plane.

Step 2 Find the lengths of the sides of each triangle. Use the Distance Formula if necessary.

$d = \sqrt{(x_2 - x_1)^2 + (y_2 - y_1)^2}$

$CD = \sqrt{(4 - 2)^2 + (-2 - 2)^2}$ $= \sqrt{4 + 16} = 2\sqrt{5}$

$FG = \sqrt{(-4 - 0)^2 + (-1 - 1)^2}$ $= \sqrt{16 + 4} = 2\sqrt{5}$

$DE = 4$

$GH = 4$

$EC = \sqrt{(2 - 0)^2 + [2 - (-2)]^2}$ $= \sqrt{4 + 16} = 2\sqrt{5}$

$HF = \sqrt{[0 - (-4)]^2 + (1 - 3)^2}$ $= \sqrt{16 + 4} = 2\sqrt{5}$

So, $\overline{CD} \cong \overline{FG}$, $\overline{DE} \cong \overline{GH}$, and $\overline{EC} \cong \overline{HF}$. Therefore $\triangle CDE \cong \triangle FGH$ by SSS, and $\angle CED \cong \angle FHG$ by CPCTC.

Use the graph to prove each congruence statement.

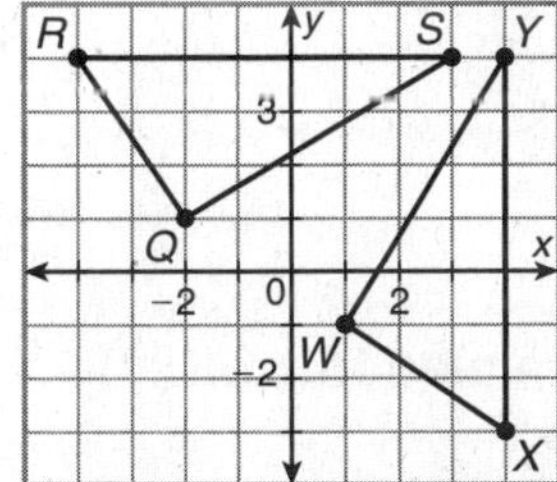

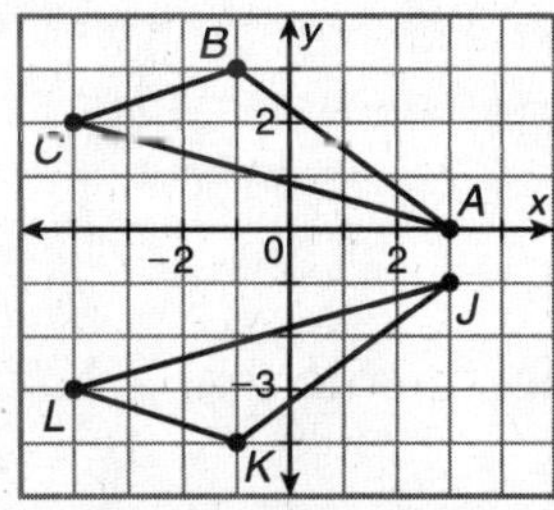

3. $\angle RSQ \cong \angle XYW$

____

____

____

____

4. $\angle CAB \cong \angle LJK$

____

____

____

____

5. Use the given set of points to prove $\angle PMN \cong \angle VTU$.

$M(-2, 4)$, $N(1, -2)$, $P(-3, -4)$, $T(-4, 1)$, $U(2, 4)$, $V(4, 0)$

____

____

Name ______________________________ Date ___________ Class ___________

LESSON 4-6

Challenge

Isosceles Triangles and Roof Trusses

The wooden or metal framework that supports a roof is called a roof *truss*. The simplest type of roof truss has the shape of an isosceles triangle, as depicted by $\triangle ABC$ at right. In the diagram, the legs of the triangle, $\overline{AB}$ and $\overline{CB}$, represent sloping beams that are called *rafters*. The base, $\overline{AC}$, represents the *tie beam* that "ties together" the rafters. However, large roofs require trusses with designs that are more complex than this.

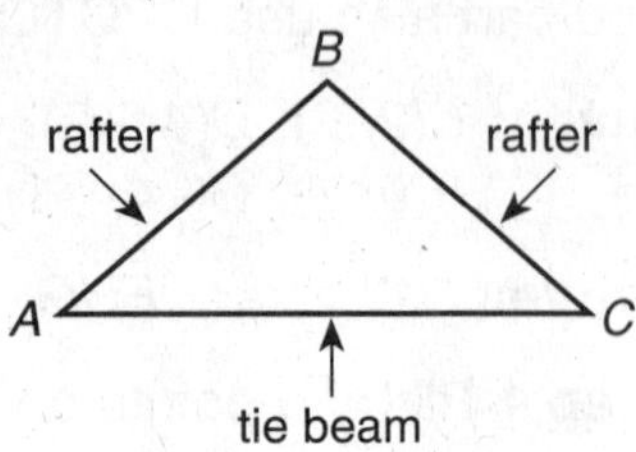

For example, a *king-post* truss is pictured at right. The king post, $\overline{KZ}$, is a median of $\triangle JKL$, and it provides support for the rafters. Additional support for the rafters comes from *struts* $\overline{ZX}$ and $\overline{ZY}$, which are medians of $\triangle JKZ$ and $\triangle LKZ$, respectively. The outer triangle, $\triangle JKL$, is an isosceles triangle with base $\overline{JL}$.

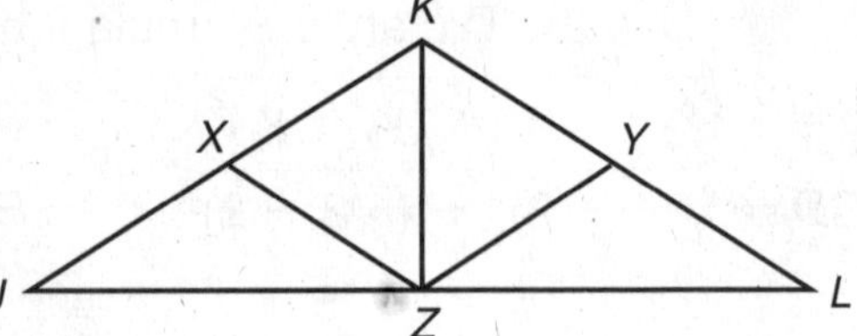

1. Refer to the diagram of the king-post truss. Write a flowchart proof to show that $\angle JKZ \cong \angle LKZ$.

2. Explain why it must be true that $\angle XJZ \cong \angle XZJ$. (*Hint:* How is $\overline{ZX}$ related to $\triangle JKL$?)

____

____

3. Given that $m\angle KJZ = 33°$, find the measures of all the other angles between the rafters, beams, and struts of the king-post truss. Label the angle measures directly on the figure.

In the *queen-post* truss pictured at right, congruent queen posts $\overline{NV}$ and $\overline{TW}$ are positioned so that they are perpendicular to tie beam $\overline{PR}$ and so that $\overline{PV} \cong \overline{VW} \cong \overline{WR}$. Struts $\overline{MV}$, $\overline{SW}$, and $\overline{NT}$ intersect the rafters so that $\overline{PM} \cong \overline{MN} \cong \overline{NQ}$ and $\overline{RS} \cong \overline{ST} \cong \overline{TQ}$. The outer triangle, $\triangle PQR$, is an isosceles triangle with base $\overline{PR}$.

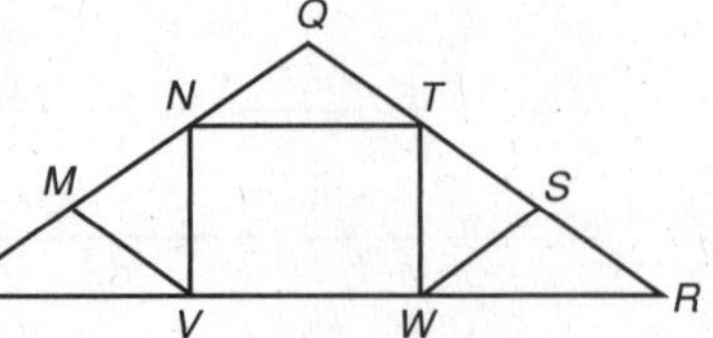

4. Using the information about the queen-post truss given above, prove each statement on a separate sheet of paper. Use any form of proof that you want.

   a. $\triangle QNT$ is isosceles. **b.** $\triangle MPV$ is isosceles. **c.** $\triangle VMN$ is isosceles.

5. Given that $m\angle NQT = 110°$, find the measures of all the other angles between the rafters, beams, and struts of the queen-post truss. Label the angle measures directly on the figure.

Name ________________________ Date ____________ Class ____________

LESSON 4-6

Problem Solving
Triangle Congruence: CPCTC

1. Two triangular plates are congruent. The area of one of the plates is 60 square inches. What is the area of the other plate? Explain.

____

____

2. An archaeologist draws the triangles to find the distance XY across a ravine. What is XY? Explain.

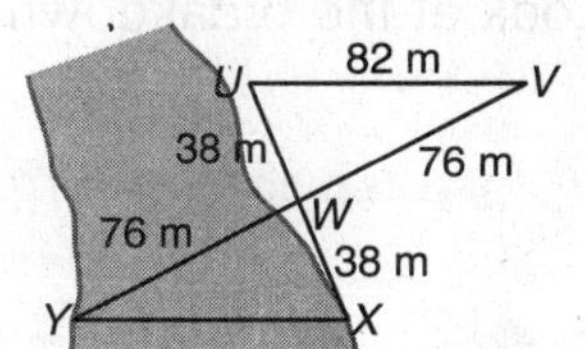

____

____

3. A city planner sets up the triangles to find the distance RS across a river. Describe the steps that she can use to find RS.

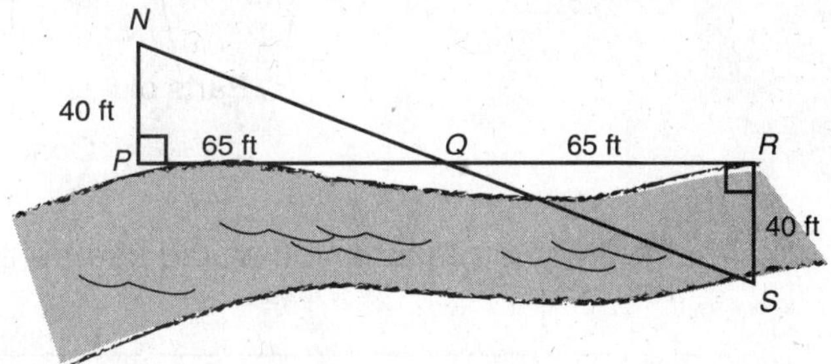

____

____

____

____

____

Choose the best answer.

4. A lighthouse and the range of its shining light are shown. What can you conclude?

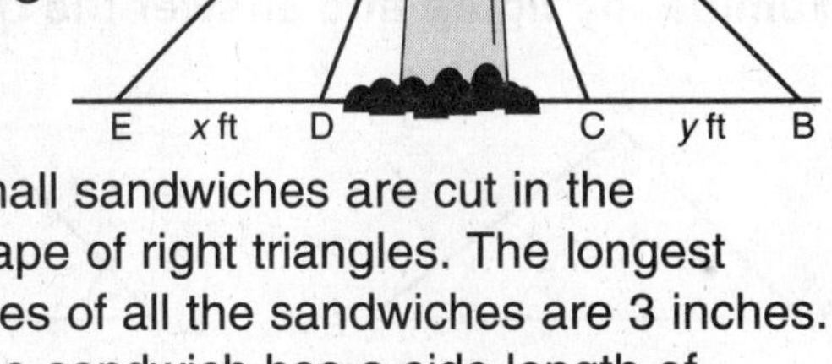

A $x = y$ by CPCTC
B $x = 2y$
C $\angle AED \cong \angle ADE$ by CPCTC
D $\angle AED \cong \angle ACB$

5. A rectangular piece of cloth 15 centimeters long is cut along a diagonal to form two triangles. One of the triangles has a side length of 9 centimeters. Which is a true statement?

F The second triangle has an angle measure of 15° by CPCTC.

G The second triangle has a side length of 9 centimeters by CPCTC.

H You cannot make a conclusion about the side length of the second triangle.

J The triangles are not congruent.

6. Small sandwiches are cut in the shape of right triangles. The longest sides of all the sandwiches are 3 inches. One sandwich has a side length of 2 inches. Which is a true statement?

A All the sandwiches have a side length of 2 inches by CPCTC.

B All the sandwiches are isosceles triangles with side lengths of 2 inches.

C None of the other sandwiches have side lengths of 2 inches.

D You cannot make a conclusion using CPCTC.

Name ______________________ Date ____________ Class ____________

LESSON 4-6

Reading Strategies

Using an Acronym

An **acronym** is a word formed from the first letters of a phrase. For example, ASAP stands for "As Soon As Possible." Acronyms can also combine the first letters or series of letters in a series of words, as in *radar*, which stands for **ra**dio **d**etecting **a**nd **r**anging.

One acronym used in geometry is CPCTC.
Look at the breakdown of this acronym:

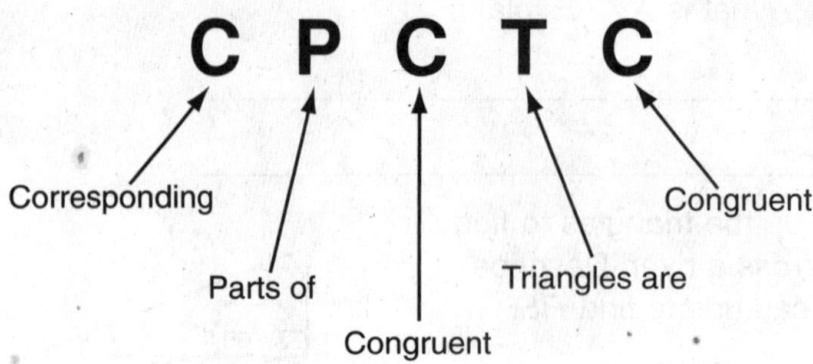

1. What are some reasons you would use an acronym?

____

____

2. What are some other acronyms you have used in your everyday life?

____

____

Examine the figure and answer the question.

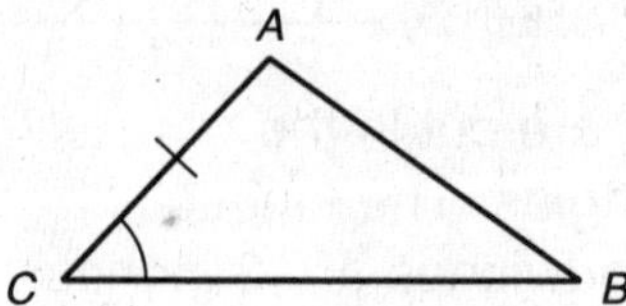

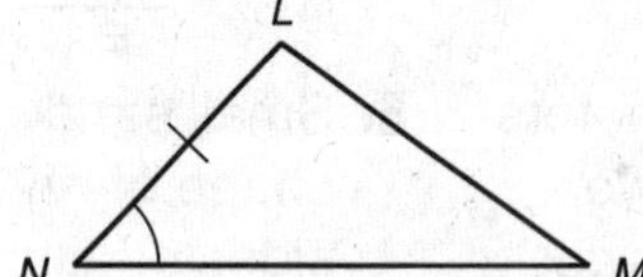

3. In this triangle, $\angle C \cong \angle N$ and $\overline{AC} \cong \overline{LN}$. Assume that $\triangle ABC \cong \triangle LMN$. Name four other parts that are congruent using CPCTC.

____

____

Name ______________________ Date __________ Class __________

LESSON 4-7

Practice A
Introduction to Coordinate Proof

Draw a square with a side length of 4 units in the coordinate plane. Label the coordinates of each vertex.

1. Center one side at the origin (0, 0).

2. Use the origin as a corner.

Complete Exercises 3–6 to finish the proof.

Given: Rectangle *ABCD* has vertices *A*(0, 4), *B*(6, 4), *C*(6, 0), and *D*(0, 0). *E* is the midpoint of $\overline{DC}$. *F* is the midpoint of $\overline{DA}$.

Prove: The perimeter of rectangle *DEGF* is one half the perimeter of rectangle *ABCD*.

3. Find the perimeter of rectangle *ABCD*. ____________

4. Use the Midpoint Formula to find the coordinates of *E* and *F*. Then find the coordinates of *G*.

E(________, ________) *F*(________, ________) *G*(________, ________)

5. Find the perimeter of rectangle *DEGF*. ____________

6. Is the perimeter of rectangle *DEGF* one-half the perimeter of rectangle *ABCD*? ____________

Complete Exercises 7–11 to finish the same proof but for all rectangles.

Given: *ABCD* is a rectangle. *E* is the midpoint of $\overline{DC}$. *F* is the midpoint of $\overline{DA}$.

Prove: The perimeter of rectangle *DEGF* is one-half the perimeter of rectangle *ABCD*.

7. Find the coordinates of *B*. (________, ________)

8. Find the perimeter of rectangle *ABCD*. ____________

9. Use the Midpoint Formula to find the coordinates of *E* and *F*. Then find the coordinates of *G*.

E(________, ________) *F*(________, ________) *G*(________, ________)

10. Find the perimeter of rectangle *DEGF*. ____________

11. Is the perimeter of rectangle *DEGF* one-half the perimeter of *ABCD*? ____________

Name ______________________ Date __________ Class __________

LESSON 4-7

Practice B

Introduction to Coordinate Proof

Position an isosceles triangle with sides of 8 units, 5 units, and 5 units in the coordinate plane. Label the coordinates of each vertex. (*Hint:* Use the Pythagorean Theorem.)

1. Center the long side on the *x*-axis at the origin.

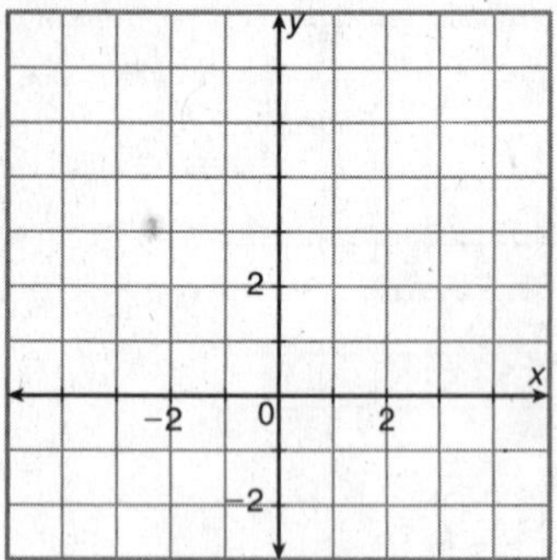

2. Place the long side on the *y*-axis centered at the origin.

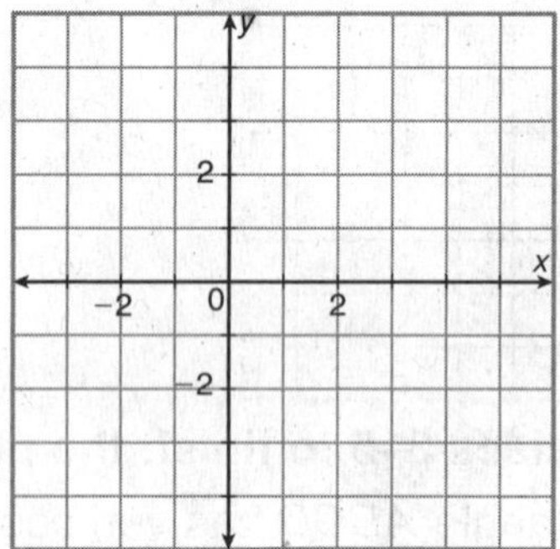

Write a coordinate proof.

3. **Given:** Rectangle *ABCD* has vertices *A*(0, 4), *B*(6, 4), *C*(6, 0), and *D*(0, 0). *E* is the midpoint of $\overline{DC}$. *F* is the midpoint of $\overline{DA}$.

   Prove: The area of rectangle *DEGF* is one-fourth the area of rectangle *ABCD*.

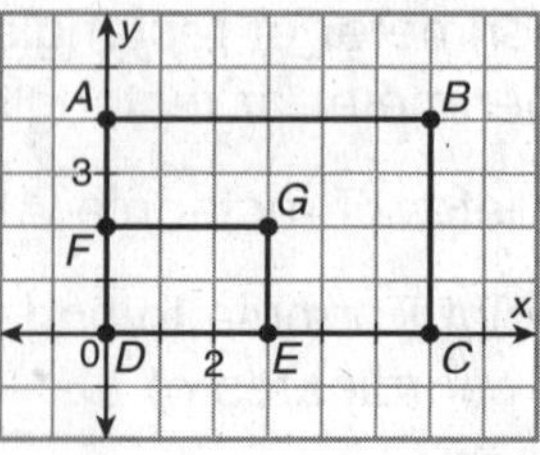

Name ____ Date ___________ Class ___________

LESSON 4-7

Practice C

Introduction to Coordinate Proof

1. A "Yield" sign is an equilateral triangle. Draw an equilateral triangle with side length ℓ in a coordinate plane so that one side is centered at the origin and falls along the *x*-axis. Determine the coordinates of each vertex in terms of ℓ.

2. A "Stop" sign is a regular octagon. A regular octagon has eight congruent sides and eight congruent 135° angles. The figure shows an octagon with side length ℓ in a coordinate plane so that one side falls along the *x*-axis and one side falls along the *y*-axis. Determine the coordinates of each vertex in terms of ℓ. (*Hint:* You will have to discover a relationship between the sides of the small right triangle at the origin.)

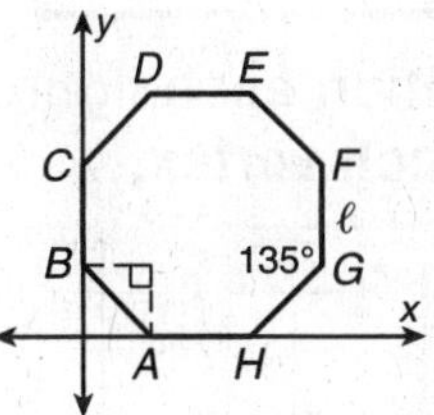

3. A "Slow down" sign is a regular hexagon. A regular hexagon has six congruent sides and six congruent 120° angles. The figure shows a hexagon with side length ℓ in a coordinate plane so that one side falls along the *x*-axis and one vertex falls on the *y*-axis. Determine the coordinates of each vertex in terms of ℓ. (*Hint:* Look back to Exercise 1.)

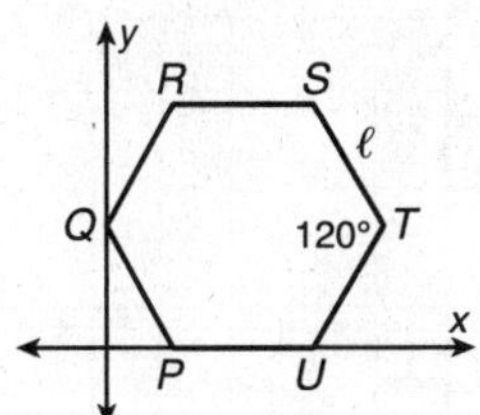

Name ______________________________ Date ____________ Class ____________

LESSON **4-7**

Reteach

Introduction to Coordinate Proof

A **coordinate proof** is a proof that uses coordinate geometry and algebra. In a coordinate proof, the first step is to position a figure in a plane. There are several ways you can do this to make your proof easier.

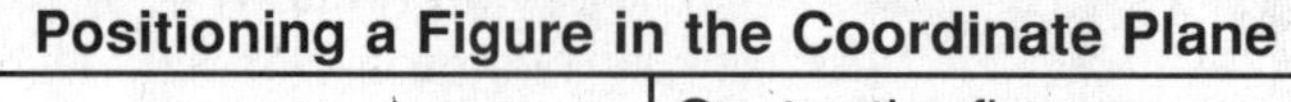

Positioning a Figure in the Coordinate Plane			
Keep the figure in Quadrant I by using the origin as a vertex.	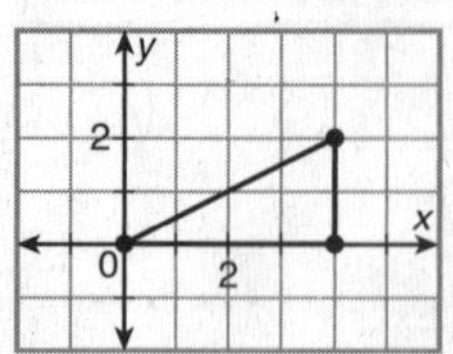	Center the figure at the origin.	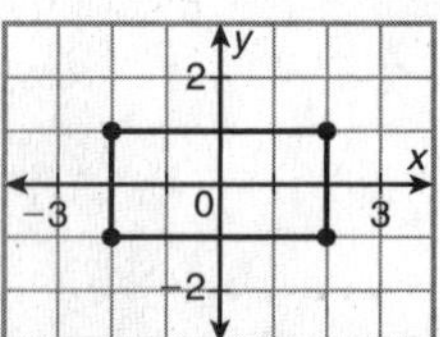
Center a side of the figure at the origin.	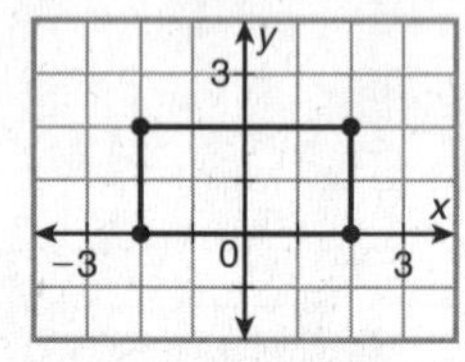	Use one or both axes as sides of the figure.	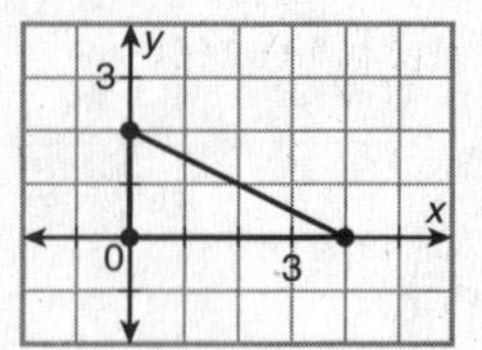

Position each figure in the coordinate plane and give the coordinates of each vertex.

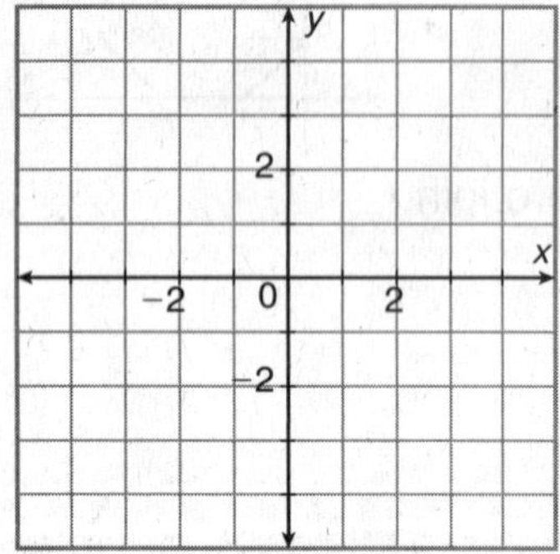

1. a square with side lengths of 6 units

____

2. a right triangle with leg lengths of 3 units and 4 units

____

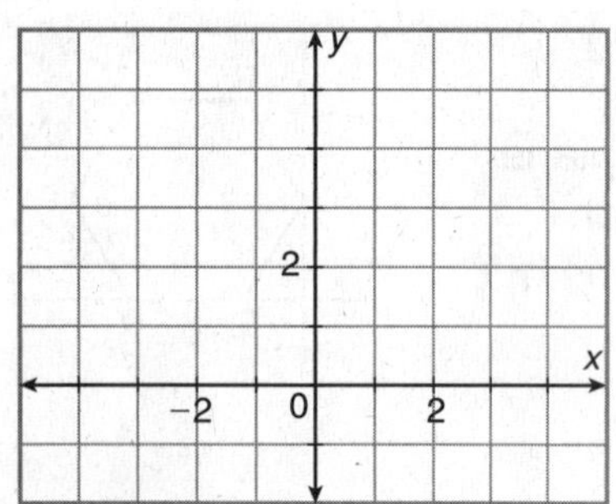

3. a triangle with a base of 8 units and a height of 2 units

____

4. a rectangle with a length of 6 units and a width of 3 units

____

Name ______________________________ Date ____________ Class ____________

LESSON 4-7

Reteach

Introduction to Coordinate Proof continued

You can prove that a statement about a figure is true without knowing the side lengths. To do this, assign variables as the coordinates of the vertices.

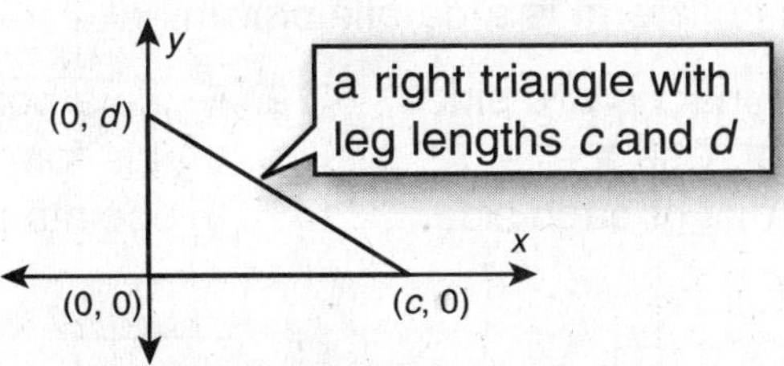

Position each figure in the coordinate plane and give the coordinates of each vertex.

5. a right triangle with leg lengths *s* and *t*

____

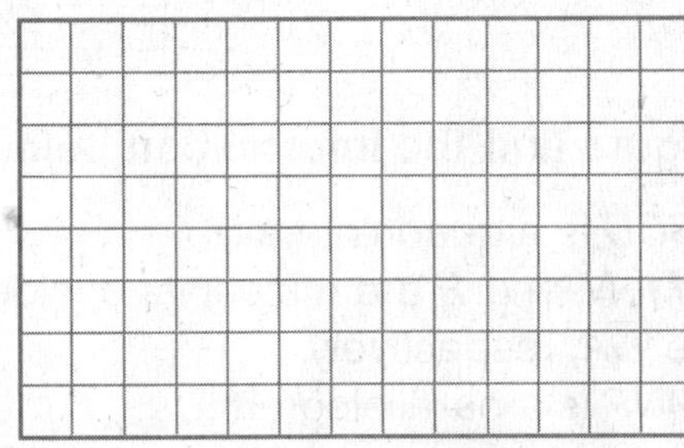

6. a square with side lengths *k*

____

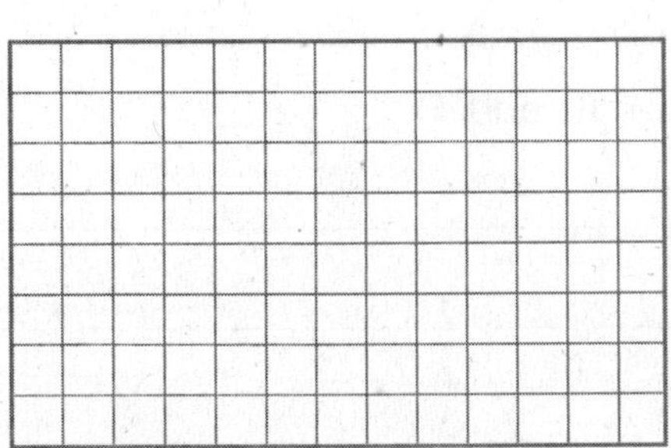

7. a rectangle with leg lengths ℓ and *w*

____

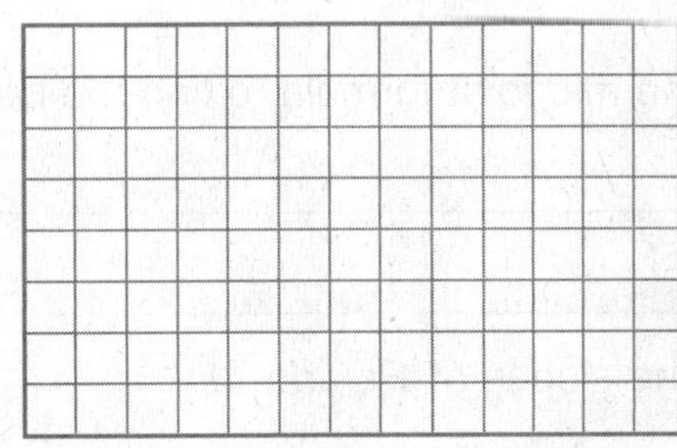

8. a triangle with base *b* and height *h*

____

9. Describe how you could use the formulas for midpoint and slope to prove the following.

Given: $\triangle HJK$, *R* is the midpoint of $\overline{HJ}$, *S* is the midpoint of $\overline{JK}$.

Prove: $\overline{RS} \parallel \overline{HK}$

____

____

____

Name ______________________ Date __________ Class __________

LESSON 4-7

Challenge

Using Coordinate Proof to Verify Conjectures

A **parallelogram** is a quadrilateral with both pairs of opposite sides parallel. If one pair of opposite sides of a quadrilateral are both parallel and congruent, then the quadrilateral is a parallelogram.

1. Draw a quadrilateral so that no two sides are congruent or parallel. Use a ruler to find and draw the midpoint of each side. Connect the midpoints.

2. Make a conjecture about the figure that is formed when the consecutive midpoints are connected.

____

____

Use the figure and the information below to answer each question.

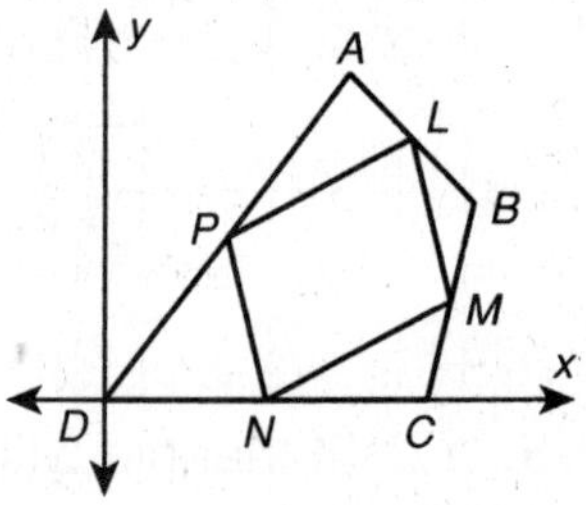

Given: *ABCD* is a quadrilateral.
L, M, N, and *P* are midpoints of sides $\overline{AB}$, $\overline{BC}$, $\overline{CD}$, and $\overline{DA}$, respectively.

Prove: *LMNP* is a parallelogram.

3. Give the coordinates of each vertex of quadrilateral *ABCD* as it is positioned in the coordinate plane. (*Hint:* Use coordinates that are multiples of 2 to make finding the midpoints easier.)

____

4. Use the midpoint formula to find the coordinates of *L, M, N,* and *P.*

____

____

5. Find the slopes of $\overline{PL}$ and $\overline{MN}$.

____

6. Use the distance formula to find *PL* and *MN.*

____

7. What can you conclude?

____

____

____

____

Name ______________________________ Date __________ Class __________

LESSON 4-7

Problem Solving

Introduction to Coordinate Proof

Round to the nearest tenth for Exercises 1 and 2.

1. A fountain is at the center of a square courtyard. If one grid unit represents one yard, what is the distance from the fountain at (0, 0) to each corner of the courtyard?

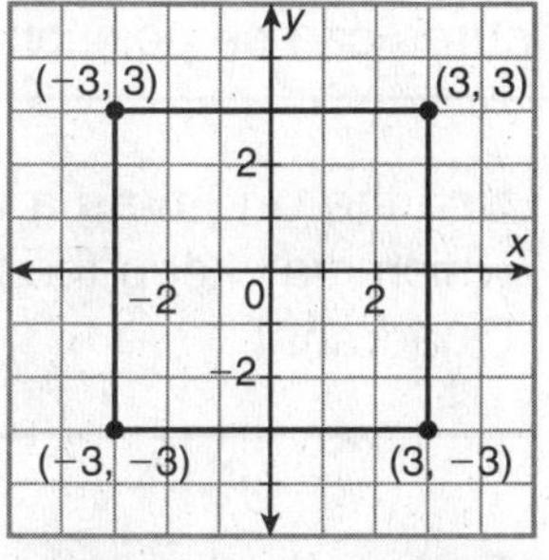

____

2. Noah started at his home at $A(0, 0)$, walked with his dog to the park at $B(4, 2)$, walked to his friend's house at $C(8, 0)$, then walked home. If one grid unit represents 20 meters, what is the distance that Noah and his dog walked?

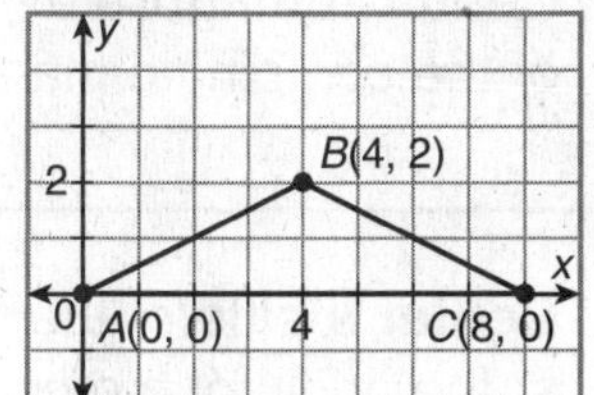

____

Use the following information for Exercises 3 and 4.

Rachel started her cycling trip at $G(0, 7)$. Malik started his trip at $J(0, 0)$. Their paths crossed at $H(4, 2)$.

3. Draw their routes in the coordinate plane.

4. If one grid unit represents $\frac{1}{2}$ mile, who had ridden farther when their paths crossed? Explain.

____

____

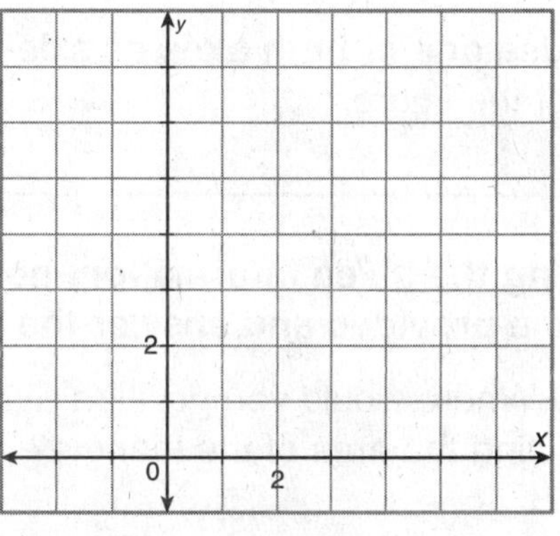

Choose the best answer.

5. Two airplanes depart from an airport at $A(9, 11)$. The first airplane travels to a location at $N(-250, 80)$, and the second airplane travels to a location at $P(105, -400)$. Each unit represents 1 mile. What is the distance, to the nearest mile, between the two airplanes?

A 335.3 mi　　**C** 490.3 mi

B 477.9 mi　　**D** 597.0 mi

6. A corner garden has vertices at $Q(0, 0)$, $R(0, 2d)$, and $S(2c, 0)$. A brick walkway runs from point Q to the midpoint M of $\overline{RS}$. What is QM?

F (c, d)　　**H** $\sqrt{c + d}$

G $c^2 + d^2$　　**J** $\sqrt{c^2 + d^2}$

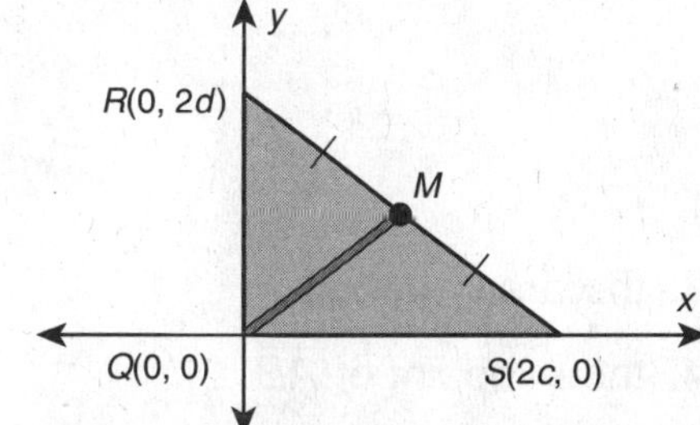

Name ______________________ Date ____________ Class ____________

LESSON 4-7

Reading Strategies

Synthesize Information

Figures can be positioned in a coordinate plane in one of four ways:

Use the **origin as a vertex,** which may keep the figure in Quadrant I.	y; (0, b); (a, b); x; (0, 0); (a, 0)
Center the figure at the origin of the coordinate plane.	y; (−a, b); (a, b); x; (−a, −b); (a, −b)
Center a side of the figure at the origin of the coordinate plane.	y; (0, b); (a, b); x; (0, −b); (a, −b)
Use **one or both axes** as sides of the figure.	y; (a, c); (a + b, c); x; (a, 0); (a + b, 0)

Using the given information, position the figure on the coordinate plane provided and answer the following questions.

1. Where would you position a triangle on a coordinate plane if you want to find the area of the triangle?

____

2. Where would you position a triangle to find the midpoint of a side?

____

Indicate where on a coordinate plane each figure should be placed in order to find the following measurement.

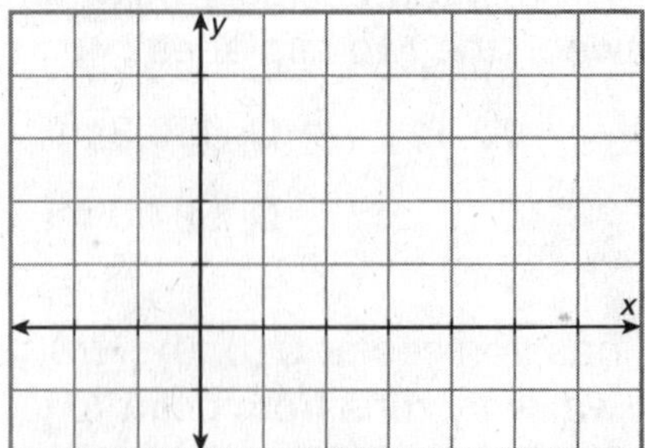

3. the area of $\triangle ABC$ ____________________

4. the midpoint of $\overline{AB}$ ____________________

5. the area of $\triangle ADC$ ____________________

6. the area of $\triangle CDB$ ____________________

Name ________________________ Date ____________ Class ____________

LESSON 4-8

Practice A

Isosceles and Equilateral Triangles

Name the parts of the figure that match the vocabulary words.

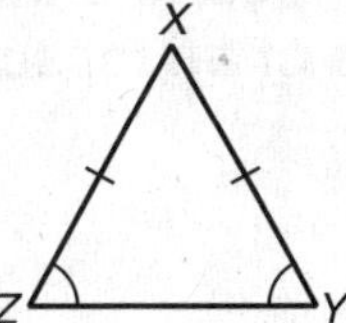

1. base: ________

2. legs: ________ and ________

3. base angles: ________ and ________

4. vertex angle: ________

Fill in the blanks in Exercises 5–8 to complete each theorem.

5. If a triangle is equilateral, then it is ________________________.

6. If two angles of a triangle are congruent, then the sides ________________________ those angles are congruent.

7. If two sides of a triangle are congruent, then the ________________________ opposite those sides are congruent.

8. If a triangle is equiangular, then it is ________________________.

9. A forest ranger in Grand Canyon National Park wants to find the minimum distance across the canyon. She finds a place in the Marble Canyon area of the park where the sides seem close together. She takes measurements and draws this figure. Find the distance *AB*. (*Hint:* The angles in an equiangular triangle measure 60°.)

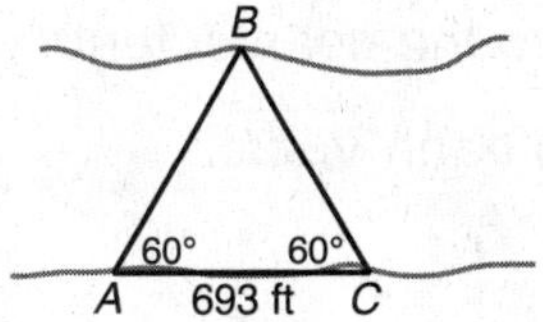

____

Find each value.

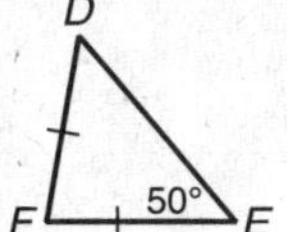

10. $m\angle D =$ ________

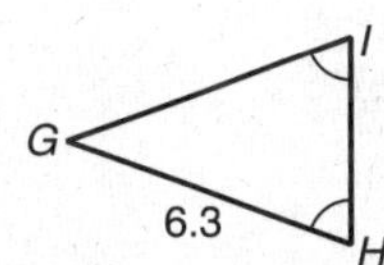

11. $GI =$ ________

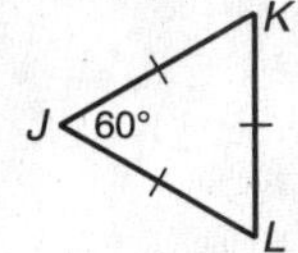

12. $m\angle L =$ ________

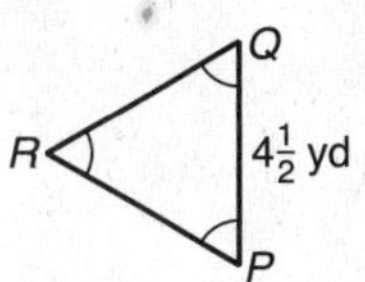

13. $RQ =$ ________

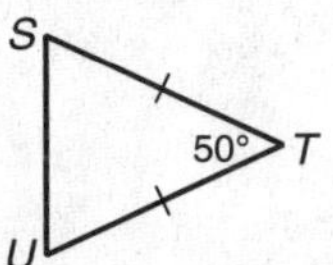

14. $m\angle U =$ ________

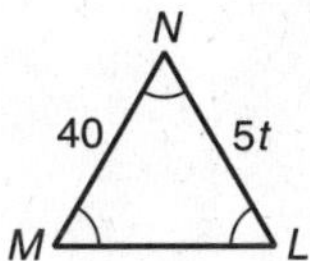

15. $t =$ ________

Name ______________________ Date ____________ Class ____________

LESSON 4-8

Practice B

Isosceles and Equilateral Triangles

An altitude of a triangle is a perpendicular segment from a vertex to the line containing the opposite side. Write a paragraph proof that the altitude to the base of an isosceles triangle bisects the base.

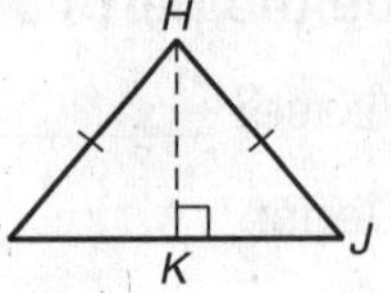

1. **Given:** $\overline{HI} \cong \overline{HJ}$, $\overline{HK} \perp \overline{IJ}$

   Prove: $\overline{HK}$ bisects $\overline{IJ}$.

2. An *obelisk* is a tall, thin, four-sided monument that tapers to a pyramidal top. The most well-known obelisk to Americans is the Washington Monument on the National Mall in Washington, D.C. Each face of the pyramidal top of the Washington Monument is an isosceles triangle. The height of each triangle is 55.5 feet, and the base of each triangle measures 34.4 feet. Find the length, to the nearest tenth of a foot, of one of the two equal legs of the triangle. ____________

Find each value.

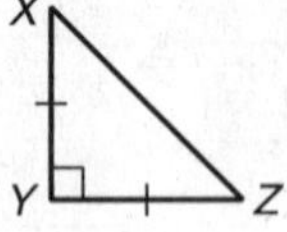

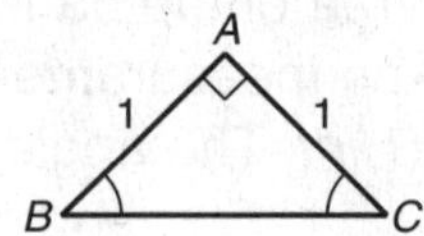

3. m∠X = ____________

4. BC = ____________

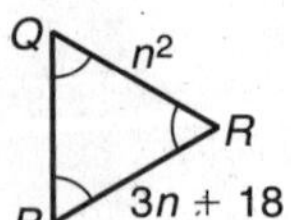

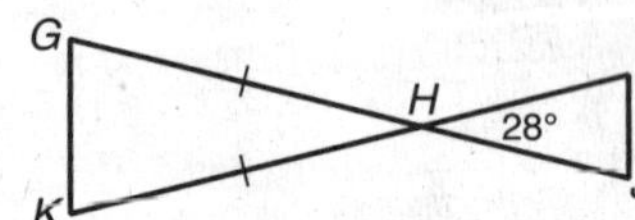

5. PQ = ____________

6. m∠K = ____________

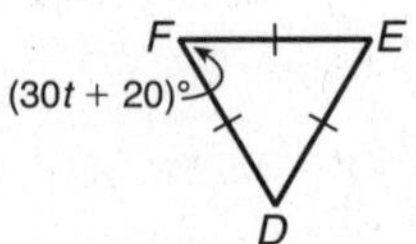

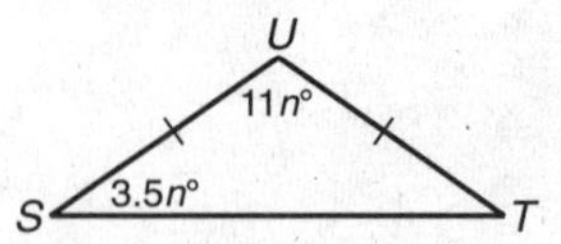

7. t = ____________

8. n = ____________

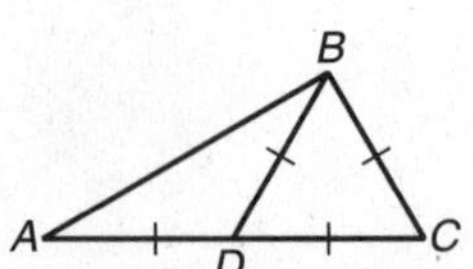

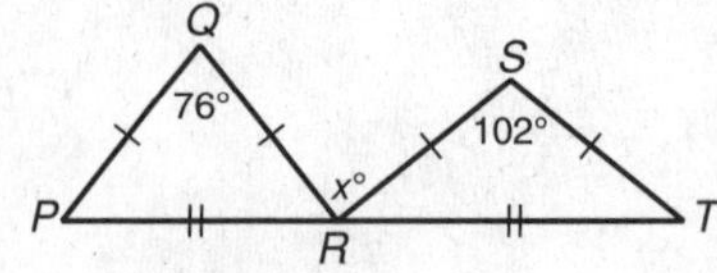

9. m∠A = ____________

10. x = ____________

Name ______________________ Date ____________ Class ____________

LESSON 4-8

Practice C
Isosceles and Equilateral Triangles

Draw a figure, name the coordinates, and write a coordinate proof.

1. **Given:** $\triangle ABC$ is an isosceles triangle. D is the midpoint of the base $\overline{BC}$.

   Prove: $\overline{AD} \perp \overline{BC}$

A soccer ball is covered by a pattern of regular pentagons and regular hexagons. On a flat surface (like this page), the edges of the shapes do not meet exactly, but on a sphere (like a ball), they cover the entire shape without any gaps. Use the figure for Exercises 2 and 3.

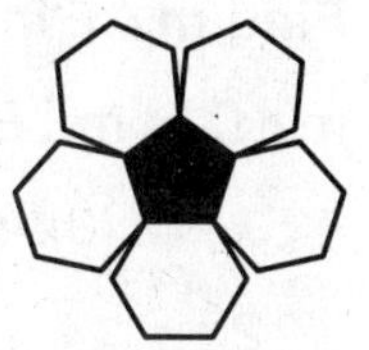

2. A regular pentagon has five congruent sides and five 108° angles. Use the figure and your knowledge of isosceles triangles to find the values of x, y, and z.

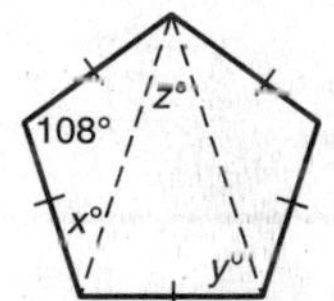

____

3. A regular hexagon has six congruent sides and six 120° angles. If a regular hexagon has side lengths of ℓ, use the figure and your knowledge of isosceles triangles to find the length of diagonal $\overline{FC}$.

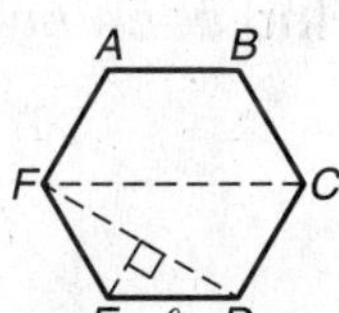

____

Name the coordinates of vertex *Y*. Write a coordinate proof.

4. **Given:** $\triangle XYZ$ is an equilateral triangle. A is the midpoint of $\overline{XY}$. B is the midpoint of $\overline{YZ}$. C is the midpoint of $\overline{XZ}$.

   Prove: $\triangle ABC$, $\triangle XAC$, $\triangle YAB$, and $\triangle CBZ$ are congruent equilateral triangles.

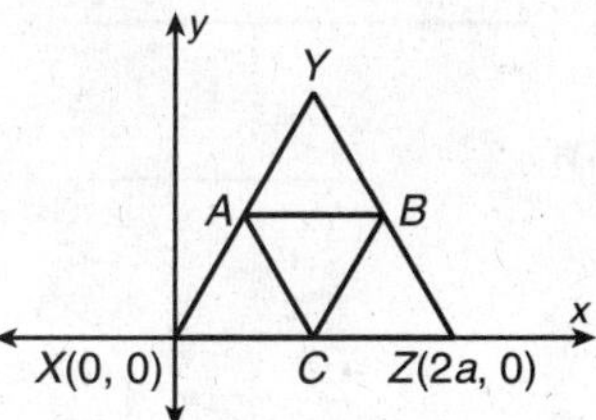

Name ______________________ Date ____________ Class ____________

LESSON 4-8

Reteach

Isosceles and Equilateral Triangles

Theorem	Examples
Isosceles Triangle Theorem If two sides of a triangle are congruent, then the angles opposite the sides are congruent.	If $\overline{RT} \cong \overline{RS}$, then $\angle T \cong \angle S$.
Converse of Isosceles Triangle Theorem If two angles of a triangle are congruent, then the sides opposite those angles are congruent.	If $\angle N \cong \angle M$, then $\overline{LN} \cong \overline{LM}$.

You can use these theorems to find angle measures in isosceles triangles.

Find m∠*E* in △*DEF*.

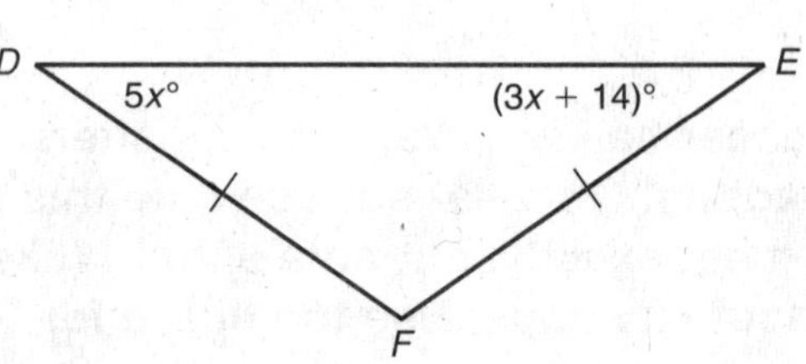

$m\angle D = m\angle E$	Isosc. △ Thm.
$5x° = (3x + 14)°$	Substitute the given values.
$2x = 14$	Subtract 3*x* from both sides.
$x = 7$	Divide both sides by 2.

Thus $m\angle E = 3(7) + 14 = 35°$.

Find each angle measure.

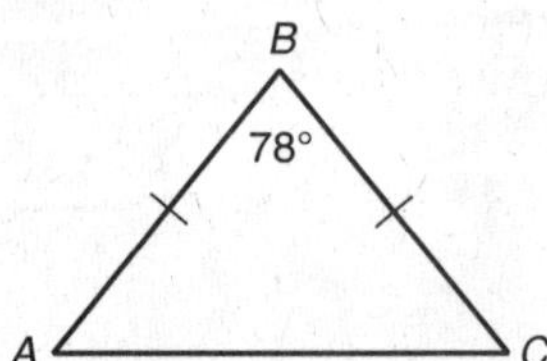

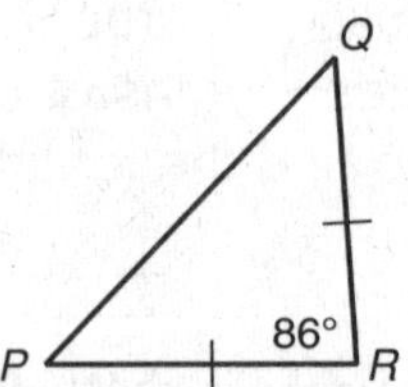

1. $m\angle C =$ ____________ **2.** $m\angle Q =$ ____________

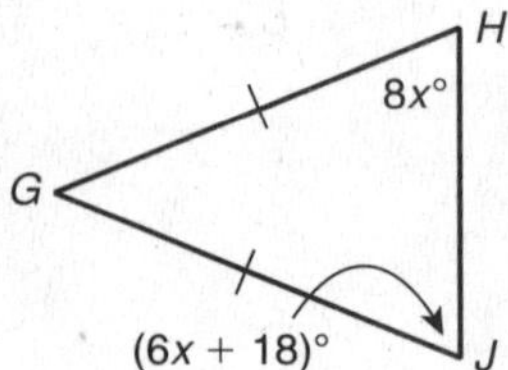

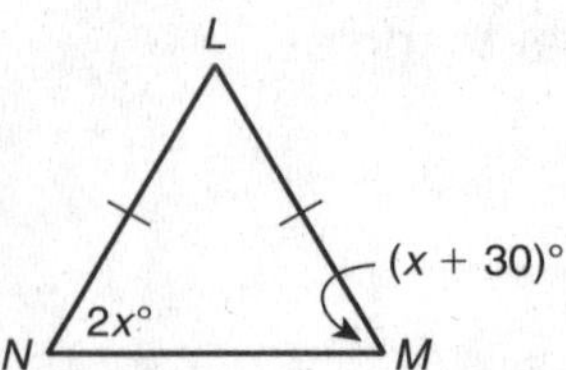

3. $m\angle H =$ ____________ **4.** $m\angle M =$ ____________

Name ______________________________ Date __________ Class __________

LESSON 4-8

Reteach

Isosceles and Equilateral Triangles continued

Equilateral Triangle Corollary
If a triangle is equilateral, then it is equiangular.

(equilateral $\triangle \rightarrow$ equiangular $\triangle$)

Equiangular Triangle Corollary
If a triangle is equiangular, then it is equilateral.

(equiangular $\triangle \rightarrow$ equilateral $\triangle$)

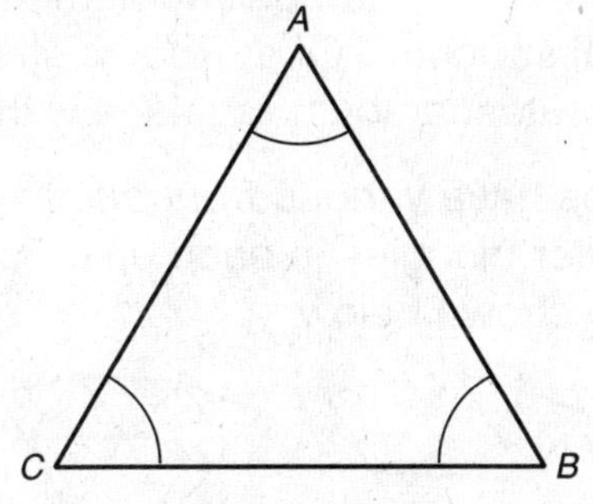

If $\angle A \cong \angle B \cong \angle C$, then $\overline{AB} \cong \overline{BC} \cong \overline{CA}$.

You can use these theorems to find values in equilateral triangles.

Find *x* in $\triangle STV$.

$\triangle STV$ is equiangular.	Equilateral $\triangle \rightarrow$ equiangular $\triangle$
$(7x + 4)° = 60°$	The measure of each $\angle$ of an equiangular $\triangle$ is 60°.
$7x = 56$	Subtract 4 from both sides.
$x = 8$	Divide both sides by 7.

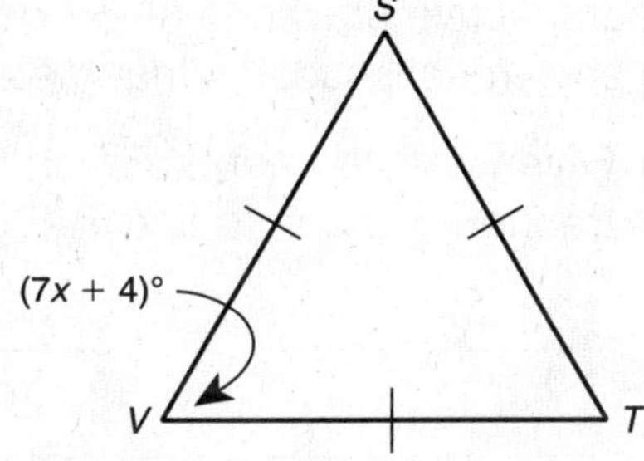

Find each value.

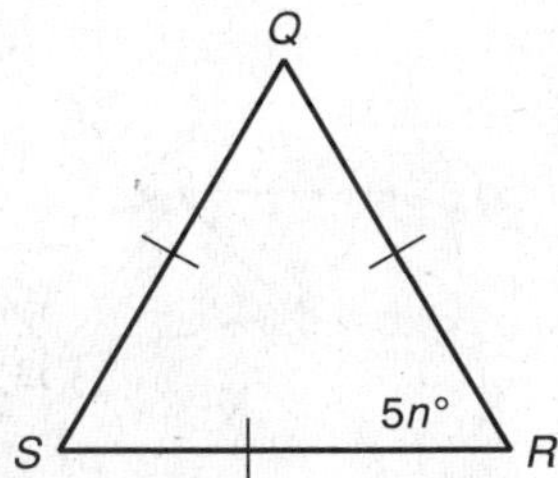

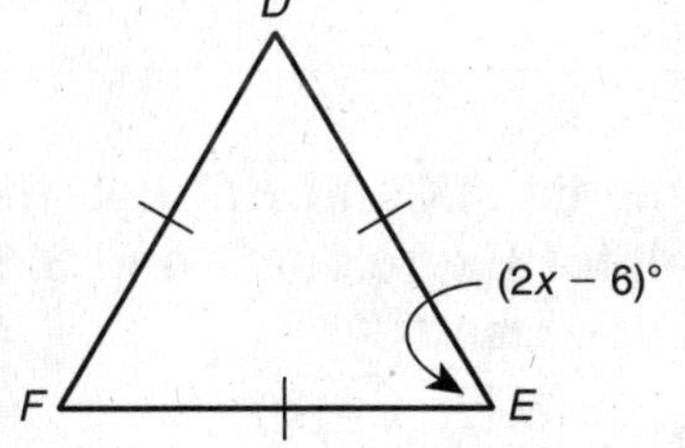

5. $n =$ ______________________

6. $x =$ ______________________

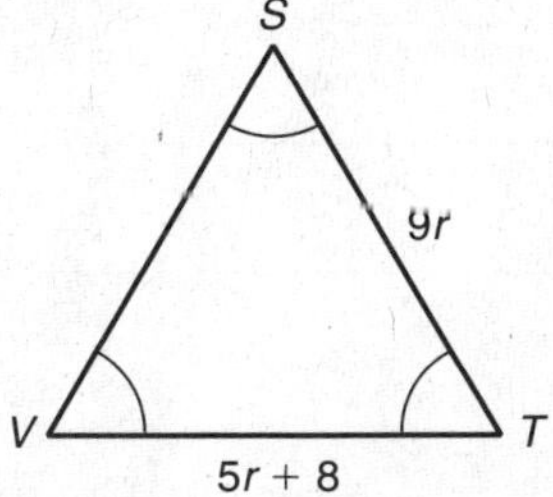

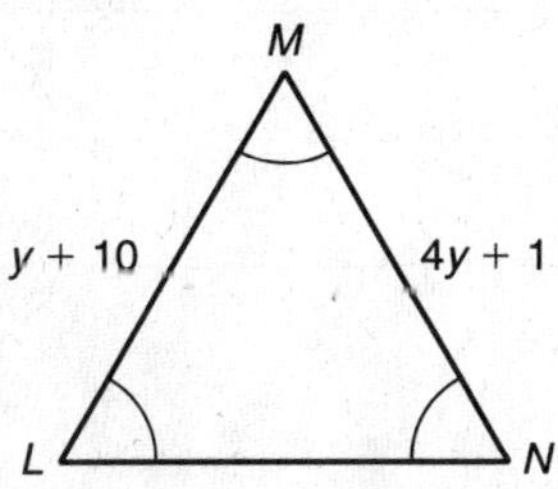

7. $VT =$ ______________________

8. $MN =$ ______________________

Name ______________________ Date __________ Class __________

LESSON 4-8

Challenge
Triangles in Geodesic Domes

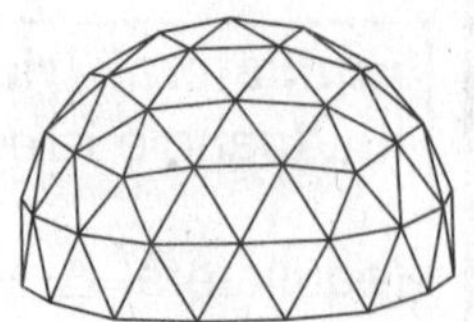

A **geodesic dome** can use equilateral triangles to create a super-strong almost spherical structure. Of all known structures, the geodesic dome encloses the greatest amount of space with the least amount of materials.

Geodesic domes have various *frequencies,* which are related to the number of smaller triangles in each unit. Four units with different frequencies are shown below.

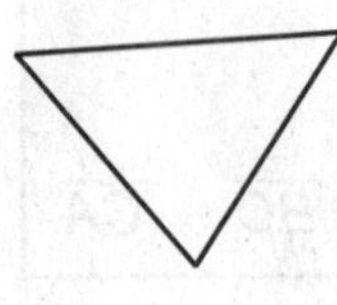

1

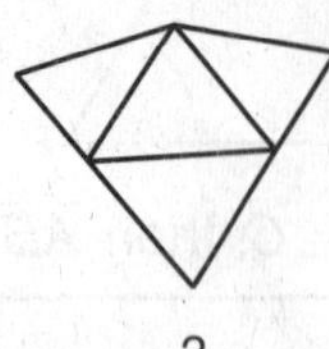

2

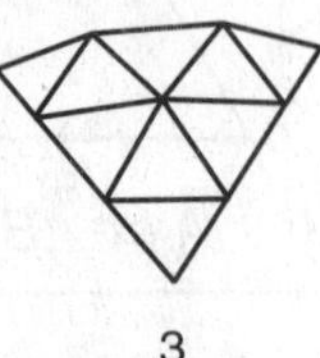

3

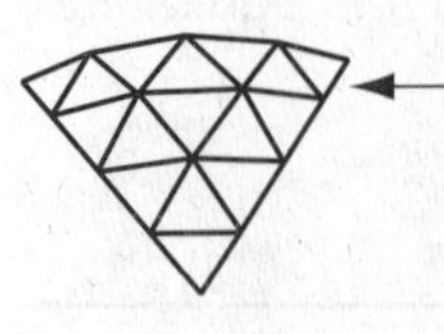

4

A high frequency dome has more triangular components and is smoother.

Frequency

In perspective drawings of geodesic domes, some of the equilateral triangles are shown as isosceles triangles.

1. Given line segments $\overline{PT}$ and $\overline{TR}$, use the diagram of the frequency 2 unit to find x. Explain your reasoning.

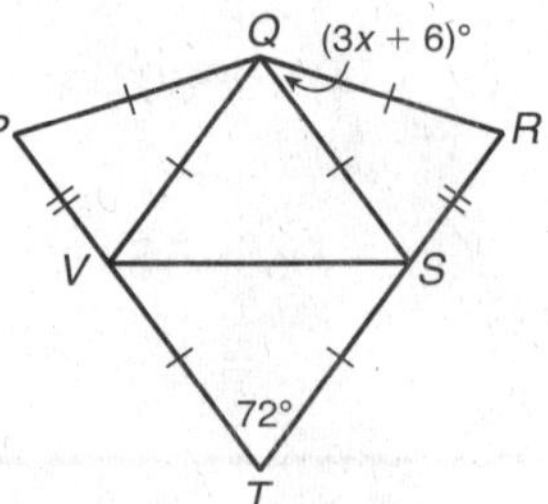

____

____

____

____

Given line segments $\overline{AK}$ and $\overline{KD}$, use the diagram of the frequency 3 unit for Exercises 2 and 3. Describe the strategies that you used.

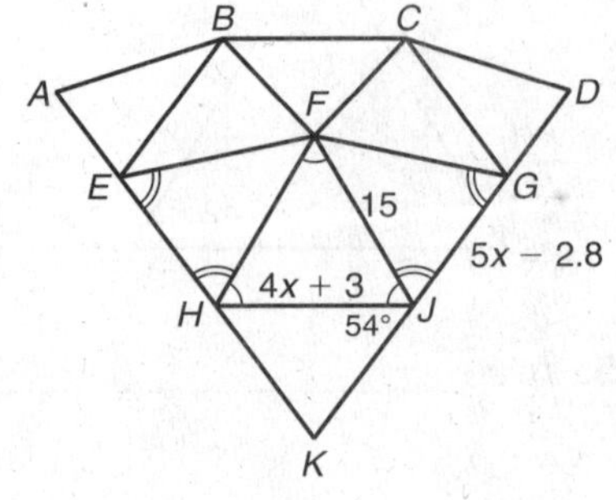

2. Find the perimeter of $\triangle EFH$.

____

____

____

____

3. Find $m\angle GFH$.

____

____

____

Name ______________________ Date ____________ Class ____________

LESSON 4-8

Problem Solving

Isosceles and Equilateral Triangles

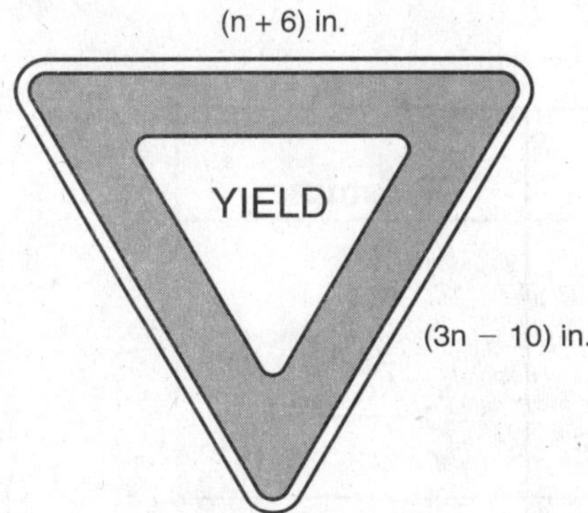

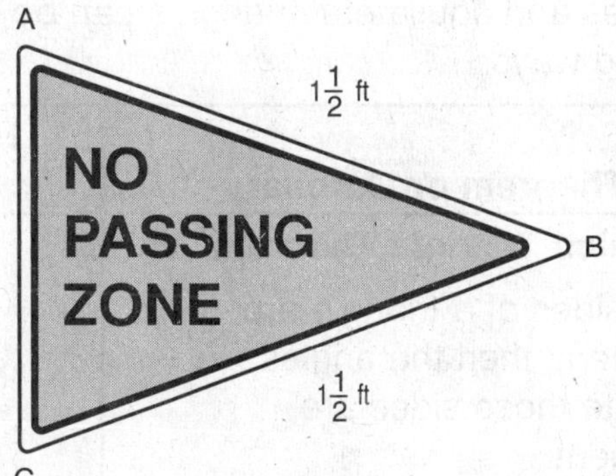

1. A "Yield" sign is an equiangular triangle. What are the lengths of the sides?

____

2. The measure of $\angle C$ is 70°. What is the measure of $\angle B$?

____

3. Samantha is swimming along $\overrightarrow{HF}$. When she is at point *H*, she sees a necklace straight ahead of her but on the bottom of the pool at point *J*. Then she swims 11 more feet to point *G*. Use the diagram to find *GJ*, the distance Samantha is from the necklace. Explain.

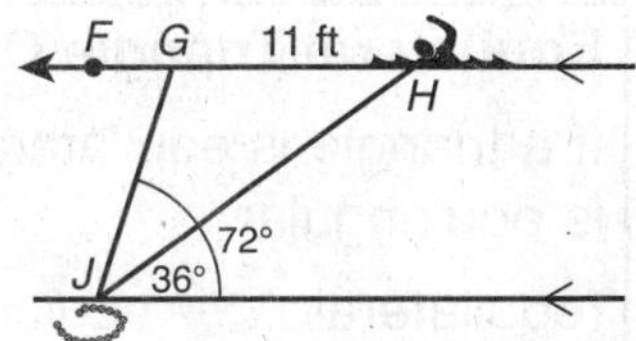

____

____

Choose the best answer.

4. A billiards triangle is equiangular. What is the perimeter?

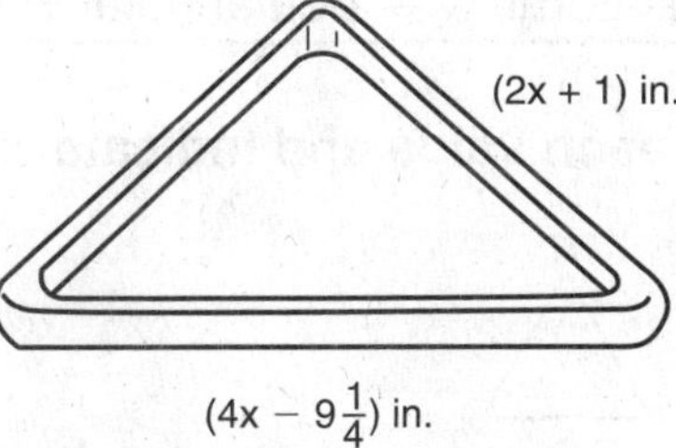

A $5\frac{1}{8}$ in. **C** $11\frac{1}{4}$ in.

B $10\frac{1}{4}$ in. **D** $33\frac{3}{4}$ in.

5. A triangular shaped trellis has angles *R*, *S*, and *T* that measure 73°, 73°, and 34°, respectively. If $ST = 4y + 6$ and $TR = 7y - 21$, what is the value of *y*?

F 5 **H** 11

G 9 **J** 15

6. Two triangular tiles each have two sides measuring 4 inches. Which is a true statement?

A Their corresponding angles are congruent.

B The triangles are congruent.

C The triangles may be congruent.

D The triangles cannot be congruent.

7. What is the value of *x* in the figure?

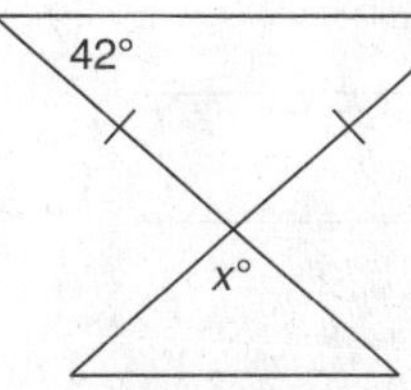

F 42° **H** 96°

G 90° **J** 106°

Name ______________________ Date ____________ Class ____________

LESSON 4-8

Reading Strategies
Understanding Relationships

Isosceles and equilateral triangles can be described in the following ways.

Theorem or Corollary	Hypothesis and Conclusion	Example
Isosceles Triangle Theorem If two sides of a triangle are congruent, then the angles opposite those sides are congruent.	If $\overline{XZ} = \overline{XY}$, then $\angle Y \cong \angle Z$.	
Converse of Isosceles Triangle Theorem If two angles of a triangle are congruent, then the sides opposite those angles are congruent.	If $\angle N \cong \angle M$, then $\overline{LM} \cong \overline{LN}$.	
Equilateral Triangle Corollary If a triangle is equilateral, then it is equiangular. (equilateral $\triangle \rightarrow$ equiangular $\triangle$)	If $\overline{QR} \cong \overline{RS} \cong \overline{SQ}$, then $\angle Q \cong \angle R \cong \angle S$.	
Equiangular Triangle Corollary If a triangle is equiangular, then it is equilateral. (equilateral $\triangle \rightarrow$ equiangular $\triangle$)	If $\angle E \cong \angle F \cong \angle G$, then $\overline{EF} \cong \overline{FG} \cong \overline{GE}$.	

Find each value and indicate which theorem you used in determining the answer.

1.

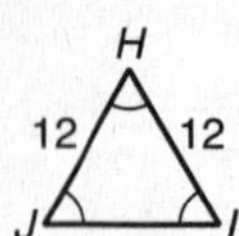

JI ______________________

____

2.

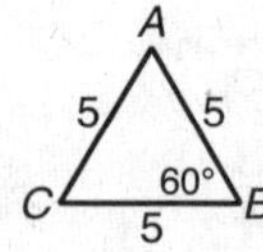

m∠*A* ______________________

____

3.

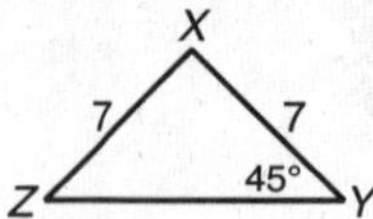

m∠*Z* ______________________

____

____

4.

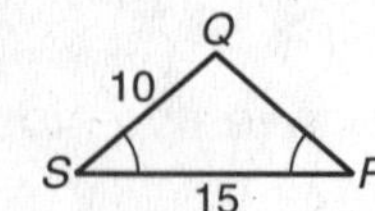

QR ______________________

____

____

LESSON 4-1

Practice A
Classifying Triangles

Match the letter of the figure to the correct vocabulary word in Exercises 1–4.

1. right triangle ___D___
2. obtuse triangle ___A___
3. acute triangle ___B, C___
4. equiangular triangle ___B___

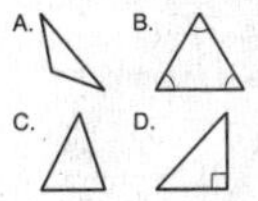

Classify each triangle by its angle measures as acute, equiangular, right, or obtuse. (*Note:* Give two classifications for Exercise 7.)

5. 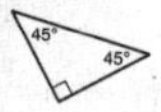 (45°, 45°) ___right___
6. 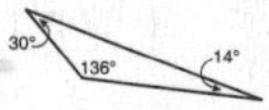 (30°, 136°, 14°) ___obtuse___
7. 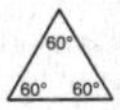 (60°, 60°, 60°) ___acute; equiangular___

For Exercises 8–10, fill in the blanks to complete each definition.

8. An isosceles triangle has ___at least two___ congruent sides.
9. An ___equilateral___ triangle has three congruent sides.
10. A ___scalene___ triangle has no congruent sides.

Classify each triangle by its side lengths as equilateral, isosceles, or scalene. (*Note:* Give two classifications in Exercise 13.)

11. 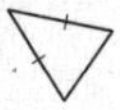 ___isosceles___
12. 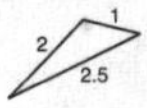 (2, 1, 2.5) ___scalene___
13.  ___isosceles; equilateral___

Find the side lengths of the triangle.

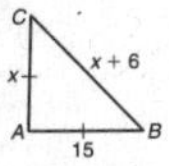

(C, A, B; x, $x + 6$, 15)

14. AB = ___15___ AC = ___15___ BC = ___21___

15. The New York City subway is known for its crowded cars. If all the seats in a car are taken, passengers must stand and steady themselves with railings or handholds. The last subway cars designed with steel hand straps were the "Redbirds" made in the late 1950s and early 1960s. The figure gives the dimensions of one of these triangular hand straps. How many hand straps could have been made from 99 inches of steel? (14 in., 14 in., 5 in.)

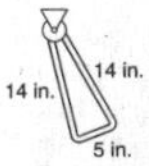

___3 hand straps___

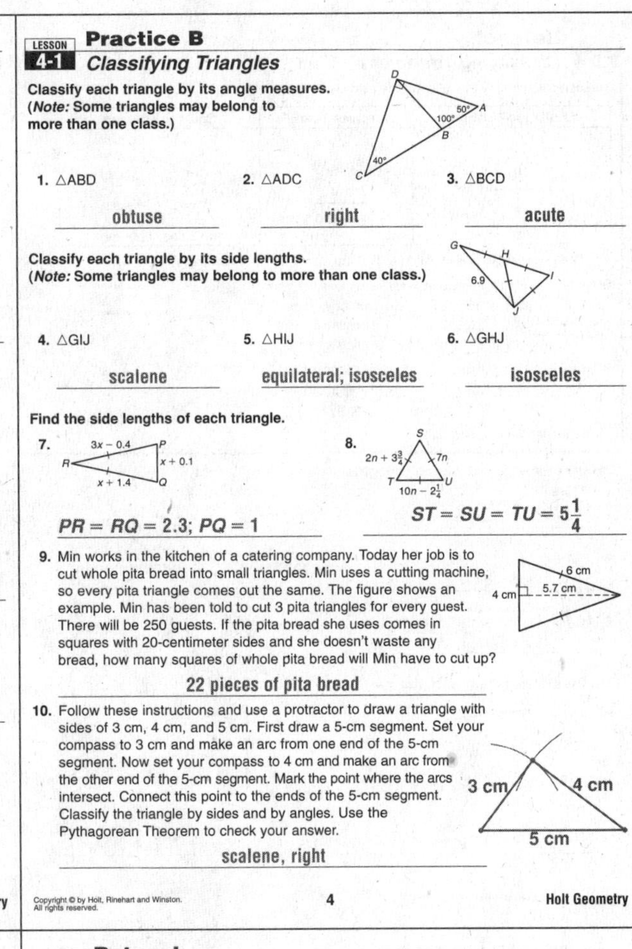

LESSON 4-1

Practice B
Classifying Triangles

Classify each triangle by its angle measures. (*Note:* Some triangles may belong to more than one class.)

(D, A, B, C; 50°, 100°, 40°)

1. △ABD ___obtuse___
2. △ADC ___right___
3. △BCD ___acute___

Classify each triangle by its side lengths. (*Note:* Some triangles may belong to more than one class.)

(G, H, I, J; 6.9)

4. △GIJ ___scalene___
5. △HIJ ___equilateral; isosceles___
6. △GHJ ___isosceles___

Find the side lengths of each triangle.

7. (R, P, Q; $3x - 0.4$, $x + 0.1$, $x + 1.4$)

$PR = RQ = 2.3$; $PQ = 1$

8. (S, T, U; $2n + 3\frac{3}{4}$, $7n$, $10n - 2\frac{1}{4}$)

$ST = SU = TU = 5\frac{1}{4}$

9. Min works in the kitchen of a catering company. Today her job is to cut whole pita bread into small triangles. Min uses a cutting machine, so every pita triangle comes out the same. The figure shows an example. Min has been told to cut 3 pita triangles for every guest. There will be 250 guests. If the pita bread she uses comes in squares with 20-centimeter sides and she doesn't waste any bread, how many squares of whole pita bread will Min have to cut up? (6 cm, 4 cm, 5.7 cm)

___22 pieces of pita bread___

10. Follow these instructions and use a protractor to draw a triangle with sides of 3 cm, 4 cm, and 5 cm. First draw a 5-cm segment. Set your compass to 3 cm and make an arc from one end of the 5-cm segment. Now set your compass to 4 cm and make an arc from the other end of the 5-cm segment. Mark the point where the arcs intersect. Connect this point to the ends of the 5-cm segment. Classify the triangle by sides and by angles. Use the Pythagorean Theorem to check your answer. (3 cm, 4 cm, 5 cm)

___scalene, right___

LESSON 4-1

Practice C
Classifying Triangles

The figure shows the side of a railway bridge. Steel girders make up the triangular pattern along the side and top. Each triangle is an equilateral triangle. In an equilateral triangle, the segment that shows the height also bisects the side it is perpendicular to. The height of the bridge is 18 feet. (18 ft)

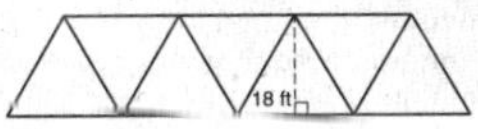

1. Find the total length of steel girder used to make this side of the bridge in feet and inches. Round to the nearest inch. ___228 ft 8 in.___
2. Find the length of wooden planks needed (that is, the span of the bridge) in feet and inches. Round to the nearest inch. ___83 ft 2 in.___

Use the figure for Exercises 3 and 4.

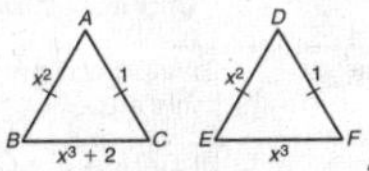

(A, B, C: x^2, 1, $x^3 + 2$; D, E, F: x^2, 1, x^3)

3. Find all possible values of x for each triangle. Explain why the solutions are different.

For △ABC, $x = 1$ or -1 because the triangles are isosceles, $x^2 = 1$, so $x = \pm 1$. For △DEF, $x \neq -1$ because a length cannot be negative, and if $x = -1$ then $EF = -1$. So $x = 1$ is the only solution for △DEF.

4. Tell what kind of triangle each must be in all cases.

△ABC must be isosceles and △DEF must be an equilateral triangle.

5. Isosceles △GHI has $\overline{GH} \cong \overline{GI}$. $GH = x^2$, $GI = -2x + 15$, and $HI = -x + 4$. Find the side lengths.

$GH = GI = 25$, $HI = 9$; $GH = GI = 9$, $HI = 1$

6. Given that $\overline{AG} \parallel \overline{BF} \parallel \overline{CE}$ and m∠A = m∠D = m∠G = 60°, write a paragraph proof proving that all the numbered angles in the figure are congruent. (Hint: Mark off each angle as you go.)

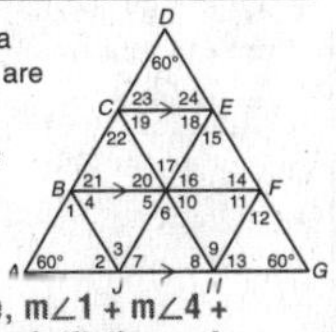

Possible answer: By the Corr. Angles Postulate, m∠A = m∠21 = m∠23 = 60° and m∠G = m∠14 = m∠24 = 60°. Construct a line parallel to $\overline{CE}$ through D. Then also by the Corr. Angles Postulate, m∠D = m∠22 = m∠1 = m∠15 = m∠12 = 60°. By the definition of a straight angle and the Angle Add. Postulate, m∠1 + m∠4 + m∠21 = 180°, but m∠1 = m∠21 = m∠A = 60°. Therefore by substitution and the Subt. Prop. of Equality, m∠4 = 60°. Similar reasoning will prove that m∠11 = m∠18 = m∠19 = 60°. By the Alt. Int. Angles Theorem, m∠19 = m∠20 and m∠18 = m∠16. m∠20 = m∠10 and m∠16 = m∠5 by the Vertical Angles Theorem. By the Alt. Int. Angles Theorem, m∠4 = m∠2 and m∠5 = m∠7 and m∠10 = m∠8 and m∠11 = m∠13. By the definition of a straight angle, the Angle Addition Postulate, substitution, and the Subt. Prop., m∠17 = m∠6 = m∠3 = m∠9. Substitution will show that the measure of every angle is 60°. Because every angle has the same measure, all of the angles are congruent by the definition of congruent angles.

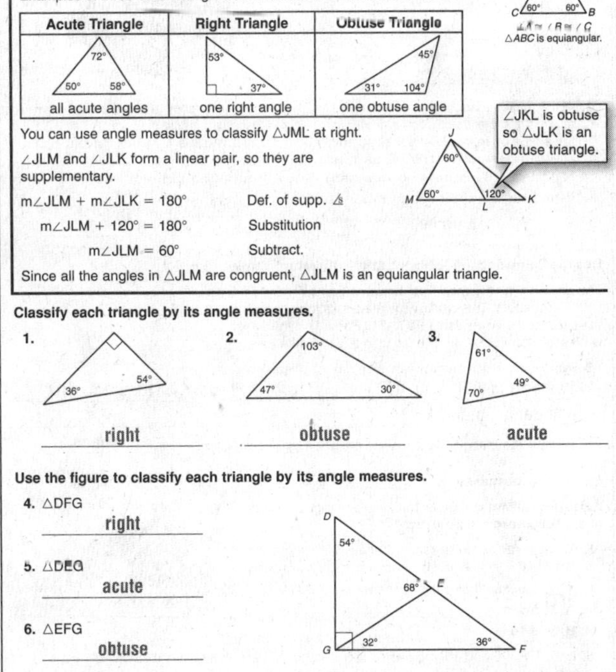

LESSON 4-1

Reteach
Classifying Triangles

You can classify triangles by their angle measures. An **equiangular triangle**, for example, is a triangle with three congruent angles.

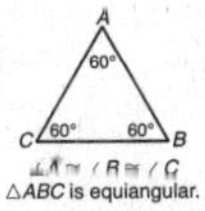

Examples of three other triangle classifications are shown in the table.

Acute Triangle	Right Triangle	Obtuse Triangle
(72°, 50°, 58°) all acute angles	(53°, 37°) one right angle	(45°, 31°, 104°) one obtuse angle

(60°, 60°, 60°) ∠A ≅ ∠B ≅ ∠C; △ABC is equiangular.

You can use angle measures to classify △JML at right.

∠JLM and ∠JLK form a linear pair, so they are supplementary.

m∠JLM + m∠JLK = 180°	Def. of supp. ∠s
m∠JLM + 120° = 180°	Substitution
m∠JLM = 60°	Subtract.

Since all the angles in △JLM are congruent, △JLM is an equiangular triangle.

∠JKL is obtuse so △JLK is an obtuse triangle.

Classify each triangle by its angle measures.

1. (36°, 54°) ___right___
2. (103°, 47°, 30°) ___obtuse___
3. (61°, 49°, 70°) ___acute___

Use the figure to classify each triangle by its angle measures.

(D, E, F, G; 54°, 68°, 32°, 36°)

4. △DFG ___right___
5. △DEG ___acute___
6. △EFG ___obtuse___

LESSON 4-1

Reteach

Classifying Triangles continued

You can also classify triangles by their side lengths.

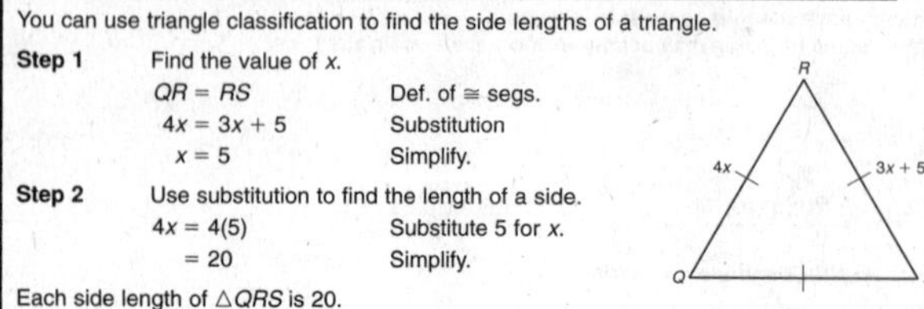

Equilateral Triangle	Isosceles Triangle	Scalene Triangle
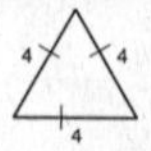 all sides congruent	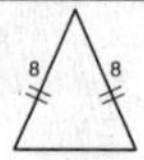 at least two sides congruent	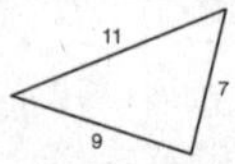 no sides congruent

You can use triangle classification to find the side lengths of a triangle.

Step 1 Find the value of x.

$QR = RS$	Def. of ≅ segs.
$4x = 3x + 5$	Substitution
$x = 5$	Simplify.

Step 2 Use substitution to find the length of a side.

$4x = 4(5)$	Substitute 5 for x.
$= 20$	Simplify.

Each side length of $\triangle QRS$ is 20.

Classify each triangle by its side lengths.

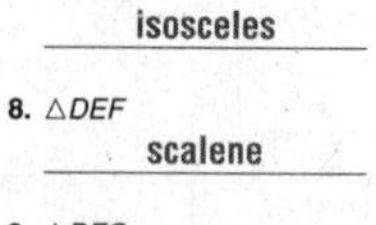

7. $\triangle EGF$

isosceles

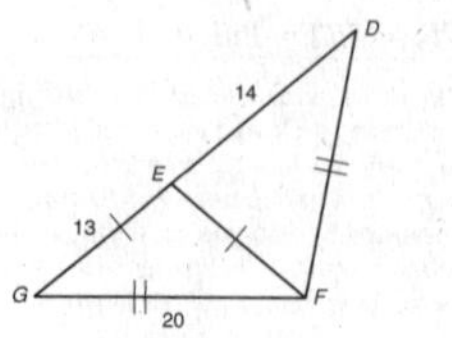

8. $\triangle DEF$

scalene

9. $\triangle DFG$

isosceles

Find the side lengths of each triangle.

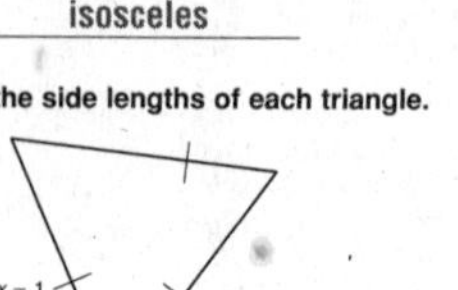

10.

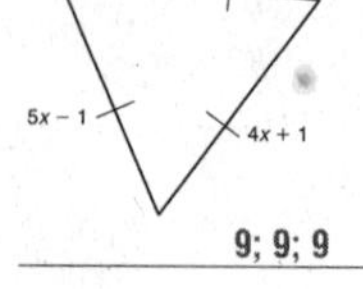

9; 9; 9

11.

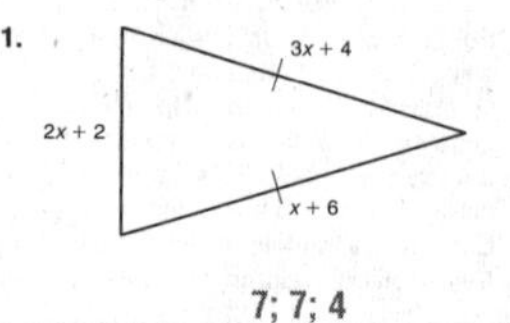

7; 7; 4

LESSON 4-1

Challenge

A Tantalizing Triangle

How many triangles are in the figure at right? Many people just count the small triangles and decide that there are sixteen in all. However, that answer accounts for only the small "one-part" triangles. Notice that there are actually four types of triangle.

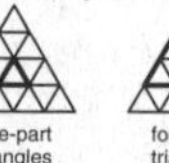

one-part triangles

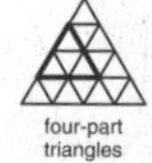

four-part triangles

nine-part triangles

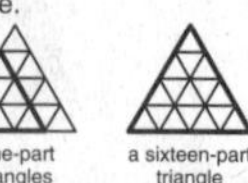

a sixteen-part triangle

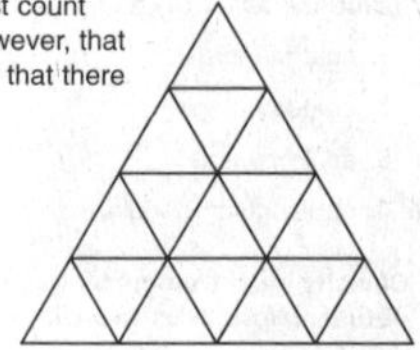

Refer to the figure at right above. How many triangles of each type are there?

1. one-part triangles 16
2. four-part triangles 7
3. nine-part triangles 3
4. sixteen-part triangles 1

5. Use your answers to Exercises 1–4. What is the total number of triangles in the figure? 27

6. What is the total number of *triangle midsegments* in the figure? 21

You may have noticed that there are several other types of geometric shapes within the figure above. For instance, the diagrams at right highlight just three of the many "three-part trapezoids" that can be found within it.

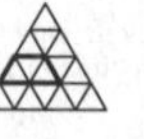

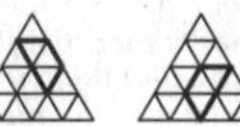

Refer to the figure above, and use the counting techniques developed in Exercises 1–6.

7. What is the total number of trapezoids in the figure? 27, 12, 3, 9, 3, 3 = 57

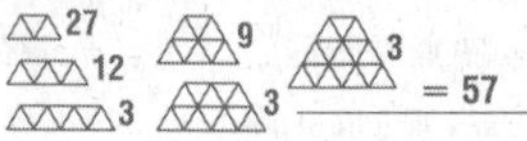

8. What is the total number of *trapezoid midsegments* in the figure? 12

9. Assume that all the small triangles are equilateral. What is the total number of rhombuses in the figure above? 18, 3 = 21

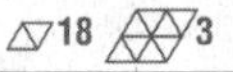

10. What is the total number of parallelograms in the figure above? 18, 3, 3, 9, 3 = 36

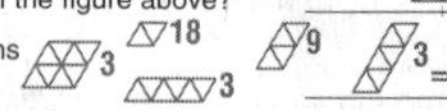

11. Identify three types of polygons that can be found in the figure above other than those named in Exercises 1–10. Your polygons may be convex or concave. Sketch each polygon in the space at right. Then find the total number of each type of polygon that you identified in the figure. 3, 3, 3, 9 Answers will vary.

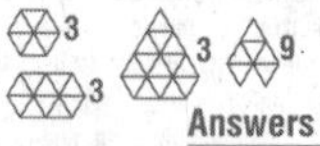

LESSON 4-1

Problem Solving

Classifying Triangles

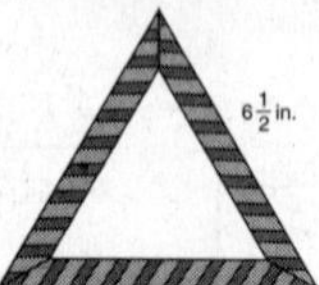

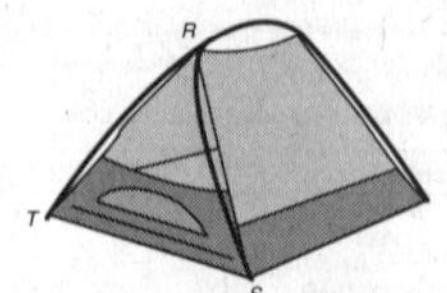

1. Aisha makes triangular picture frames by gluing three pieces of wood together in the shape of an equilateral triangle and covering the wood with ribbon. Each side of a frame is $6\frac{1}{2}$ inches long. How many frames can she cover with 2 yards of ribbon?

3 frames

2. A tent's entrance is in the shape of an isosceles triangle in which $\overline{RT} \cong \overline{RS}$. The length of $\overline{TS}$ is 1.2 times the length of a side. The perimeter of the entrance is 14 feet. Find each side length.

$4\frac{3}{8}$ ft; $4\frac{3}{8}$ ft; $5\frac{1}{4}$ ft

Use the figure and the following information for Exercises 3 and 4.

The distance "as the crow flies" between Santa Fe and Phoenix is 609 kilometers. This is 245 kilometers less than twice the distance between Santa Fe and El Paso. Phoenix is 48 kilometers closer to El Paso than it is to Santa Fe.

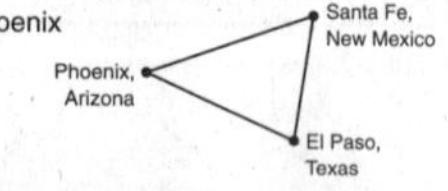

3. What is the distance between each pair of cities?

Santa Fe and El Paso, 427 km; El Paso and Phoenix, 561 km; Phoenix and Santa Fe, 609 km

4. Classify the triangle that connects the cities by its side lengths. scalene

Choose the best answer.

A *gable*, as shown in the diagram, is the triangular portion of a wall between a sloping roof.

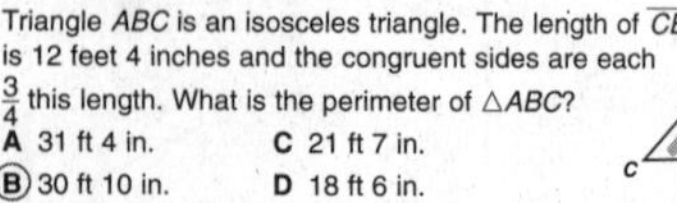

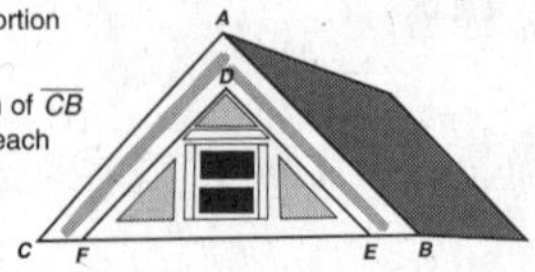

5. Triangle ABC is an isosceles triangle. The length of $\overline{CB}$ is 12 feet 4 inches and the congruent sides are each $\frac{3}{4}$ this length. What is the perimeter of $\triangle ABC$?

A 31 ft 4 in.
(B) 30 ft 10 in.
C 21 ft 7 in.
D 18 ft 6 in.

6. In $\triangle DEF$, $\overline{DE}$ and $\overline{DF}$ are each 6 feet 3 inches long. This length is 0.75 times the length of $\overline{FE}$. What is the perimeter of $\triangle DEF$?

F 12 ft 4 in.
G 14 ft 7 in.
H 17 ft 2 in.
(J) 20 ft 10 in.

LESSON 4-1

Reading Strategies

Vocabulary Development

The table below shows seven ways to classify different triangles, by angle measures and by side lengths. Remember, you cannot simply assume things about segment lengths and angle measures. Information must be given in writing or by marks and labels in the diagram.

Classification	Description	Example
acute triangle	triangle that has *three* acute **angles**	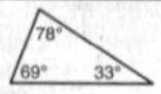 78°, 69°, 33°
equiangular triangle	triangle that has *three* congruent acute **angles**	
right triangle	triangle that has *one* right **angle**	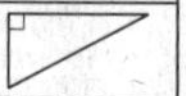
obtuse triangle	triangle that has *one* obtuse **angle**	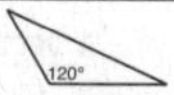 120°
equilateral triangle	triangle with *three* congruent **sides**	
isosceles triangle	triangle that has *at least two* congruent **sides**	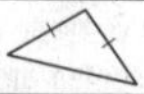
scalene triangle	triangle that has *no* congruent **sides**	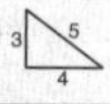 3, 5, 4

Classify the following triangles by side lengths and angle measures. There will be more than one answer.

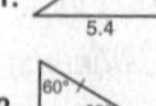

1. 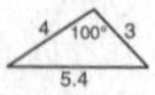 scalene, obtuse

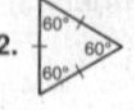

2. equilateral, equiangular, acute

3. isosceles, right

Suppose you are asked to draw a triangle by using the given information. If you think it is possible to draw such a triangle, classify the triangle. Otherwise write *no such triangle*.

4. $\triangle ABC$ with $AB = 3$, $BC = 3$, and $CA = 5$ isosceles triangle
5. $\triangle XOZ$ with $m\angle X = 92°$, $m\angle O = 92°$, and $m\angle Z = 27°$ no such triangle
6. $\triangle MNK$ with $m\angle M = 90°$, $m\angle N = 60°$, and $m\angle K = 30°$ scalene right triangle

LESSON 4-2
Practice A
Angle Relationships in Triangles

Use the figure for Exercises 1–3. Name all the angles that fit the definition of each vocabulary word.

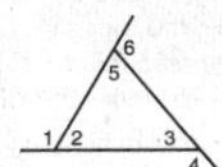

1. exterior angle ∠1, ∠4, ∠6
2. remote interior angles to ∠6 ∠2, ∠3
3. interior angle ∠2, ∠3, ∠5

For Exercises 4–7, fill in the blanks to complete each theorem or corollary.

4. The measure of each angle of an **equiangular** triangle is 60°.
5. The sum of the angle measures of a triangle is **180°**.
6. The acute angles of a **right** triangle are complementary.
7. The measure of an **exterior angle** of a triangle is equal to the sum of the measures of its remote interior angles.

Find the measure of each angle.

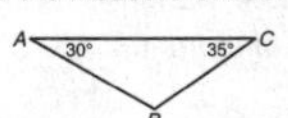

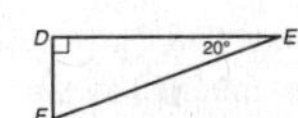

8. m∠B **115°**
9. m∠F **70°**

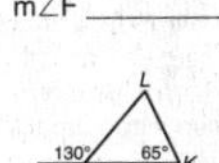

10. m∠G **60°**
11. m∠L **65°**

12. m∠P **35°**
13. m∠VWY **120°**

14. When a person's joint is injured, the person often goes through rehabilitation under the supervision of a doctor or physical therapist to make sure the joint heals well. Rehabilitation involves stretching and exercises. The figure shows a leg bending at the knee during a rehabilitation session. Use what you know about triangles to find the angle measure that the knee is bent from the horizontal (fully extended) position.

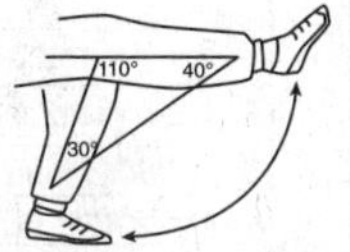

110°

LESSON 4-2
Practice B
Angle Relationships in Triangles

1. An area in central North Carolina is known as the Research Triangle because of the relatively large number of high-tech companies and research universities located there. Duke University, the University of North Carolina at Chapel Hill, and North Carolina State University are all within this area. The Research Triangle is roughly bounded by the cities of Chapel Hill, Durham, and Raleigh. From Chapel Hill, the angle between Durham and Raleigh measures 54.8°. From Raleigh, the angle between Chapel Hill and Durham measures 24.1°. Find the angle between Chapel Hill and Raleigh from Durham. **101.1°**

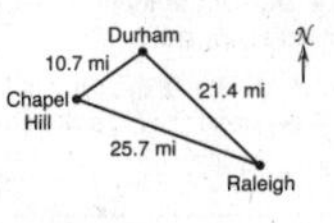

2. The acute angles of right triangle ABC are congruent. Find their measures. **45°**

The measure of one of the acute angles in a right triangle is given. Find the measure of the other acute angle.

3. 44.9° **45.1°**
4. $(90 - z)°$ **$z°$**
5. 0.3° **89.7°**

Find each angle measure.

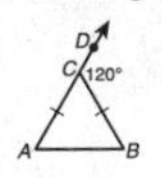

6. m∠B **60°**
7. m∠PRS **47°**

8. In △LMN, the measure of an exterior angle at N measures 99°. $m\angle L = \frac{1}{3}x°$ and $m\angle M = \frac{2}{3}x°$. Find m∠L, m∠M, and m∠LNM. **33°; 66°; 81°**
9. m∠E and m∠G **44°; 44°**
10. m∠T and m∠V **108°; 108°**

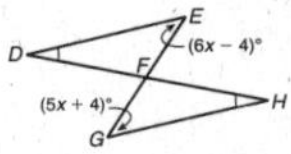

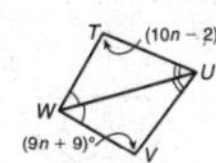

11. In △ABC and △DEF, m∠A = m∠D and m∠B = m∠E. Find m∠F if an exterior angle at A measures 107°, $m\angle B = (5x + 2)°$, and $m\angle C = (5x + 5)°$. **55°**
12. The angle measures of a triangle are in the ratio 3 : 4 : 3. Find the angle measures of the triangle. **54°; 72°; 54°**

LESSON 4-2
Practice C
Angle Relationships in Triangles

1. Write a two-column proof that the sum of the angle measures of a quadrilateral is 360°. Begin by drawing quadrilateral ABCD. (Hint: You will have to draw one auxiliary line.)

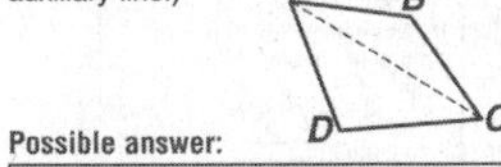

Possible answer:

Statements	Reasons
1. Quadrilateral *ABCD*	1. Given
2. Draw $\overline{AC}$.	2. Construction
3. m∠D + m∠DAC + m∠DCA = 180°, m∠B + m∠BAC + m∠BCA = 180°	3. Triangle Sum Thm.
4. m∠D + m∠DAC + m∠DCA + m∠B + m∠BAC + m∠BCA = 360°	4. Add. Prop. of =
5. m∠DAC + m∠BAC = m∠DAB, m∠DCA + m∠BCA = m∠DCB	5. Angle Add. Post.
6. m∠D + m∠DAB + m∠B + m∠DCB = 360°	6. Subst.

2. Use this figure to write a flowchart proof that the sum of the measures of the exterior angles of a triangle, one at each vertex, is 360°.

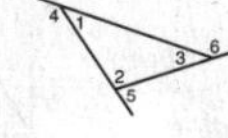

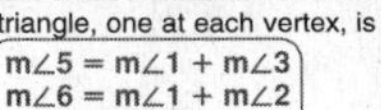

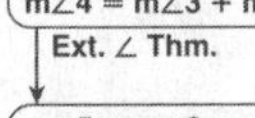

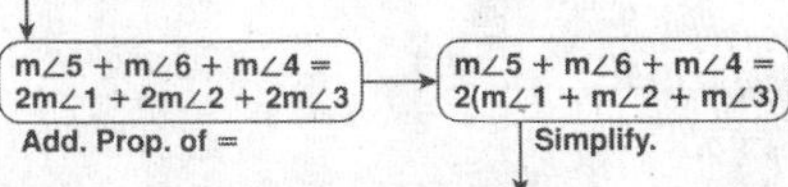

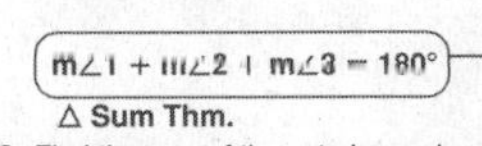

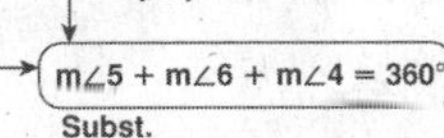

3. Find the sum of the exterior angles, one at each vertex, of a quadrilateral. **360°**
4. Use the techniques you developed in Exercises 1–3 to find the sums of the measures of the interior angles and of the exterior angles, one at each vertex, of a pentagon. **interior = 540°; exterior = 360°**
5. A landscape artist plans to draw a pair of mountains. He wants his drawing to be reasonably accurate, so he takes some measurements and draws this figure. Find x, y, and z. **$x = 93$; $y = 52$; $z = 35$**

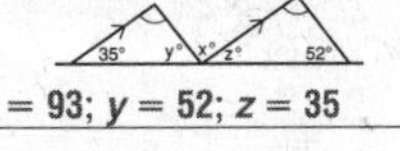

LESSON 4-2
Reteach
Angle Relationships in Triangles

According to the **Triangle Sum Theorem,** the sum of the angle measures of a triangle is 180°.

m∠J + m∠K + m∠L = 62 + 73 + 45

= 180°

The **corollary** below follows directly from the Triangle Sum Theorem.

Corollary	Example
The acute angles of a right triangle are complementary.	m∠C = 90 − 39 = 51° m∠C + m∠E = 90°

Use the figure for Exercises 1 and 2.

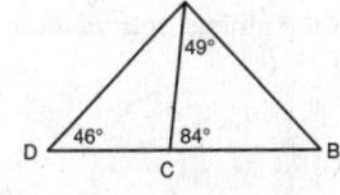

1. Find m∠ABC. **47°**

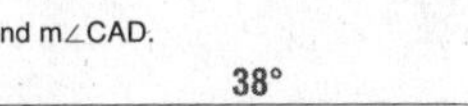

2. Find m∠CAD. **38°**

Use △*RST* for Exercises 3 and 4.

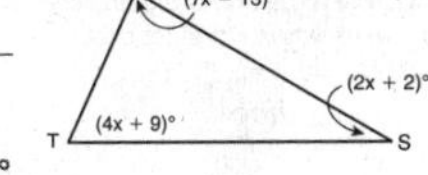

3. What is the value of x? **14**
4. What is the measure of each angle? **m∠*R* = 85°; m∠*S* = 30°; m∠*T* = 65°**

What is the measure of each angle?

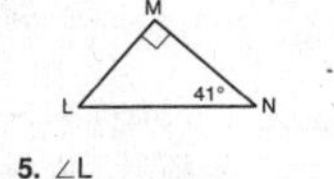

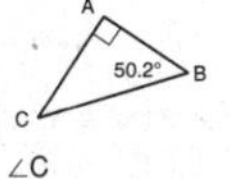

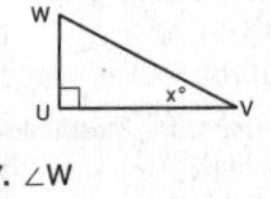

5. ∠L **49°**
6. ∠C **39.8°**
7. ∠W **$(90 - x)°$**

LESSON 4-2

Reteach

Angle Relationships in Triangles continued

An **exterior angle** of a triangle is formed by one side of the triangle and the extension of an adjacent side.

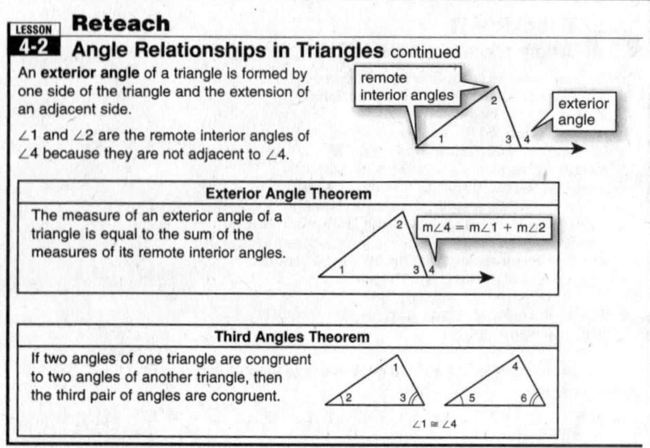

∠1 and ∠2 are the remote interior angles of ∠4 because they are not adjacent to ∠4.

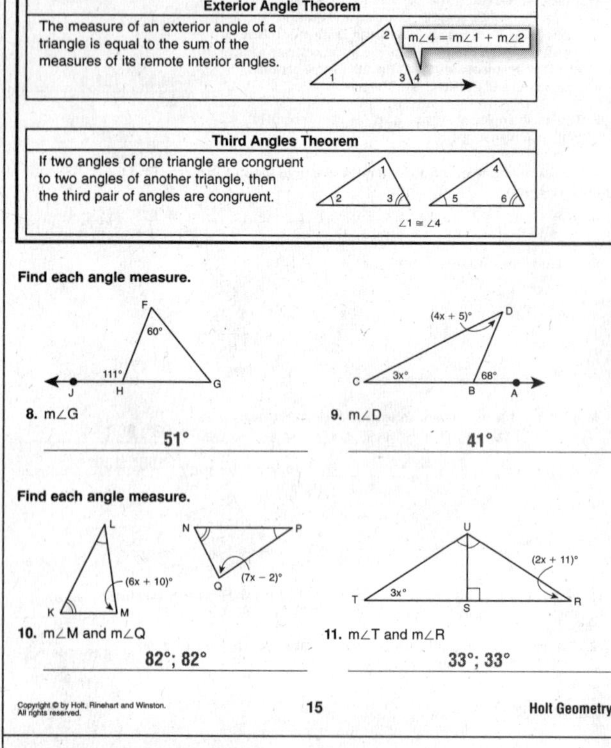

Exterior Angle Theorem

The measure of an exterior angle of a triangle is equal to the sum of the measures of its remote interior angles.

m∠4 = m∠1 + m∠2

Third Angles Theorem

If two angles of one triangle are congruent to two angles of another triangle, then the third pair of angles are congruent.

∠1 ≅ ∠4

Find each angle measure.

8. m∠G

51°

9. m∠D

41°

Find each angle measure.

10. m∠M and m∠Q

82°; 82°

11. m∠T and m∠R

33°; 33°

LESSON 4-2

Challenge

Analyzing Verbal Descriptions and Using Auxiliary Figures

Find the measure of each angle.

1. In △FGH, the measure of ∠H is 14° less than the measure of ∠F. The measure of ∠G is 25° more than $2\frac{1}{3}$ times the measure of ∠F. Find m∠F.

39°

2. In △WXY, the measure of ∠X is 4° less than twice the measure of ∠W. The measure of ∠Y is 24° less than $3\frac{1}{2}$ times m∠W. Find m∠Y.

88°

$\overline{BA} \parallel \overline{DE}$. Explain how you could draw one or more auxiliary figures to help you find the value of *x*. Then find the value of *x*. Explain.

3.

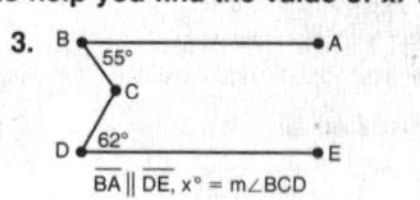

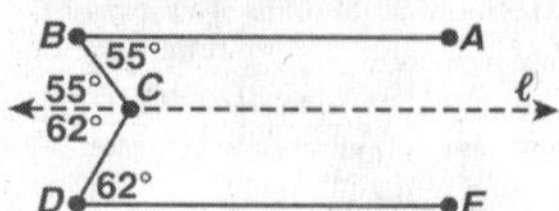

Through *C*, draw line ℓ ∥ $\overline{BA}$. So ℓ is also ∥ *DE*. Then apply the Alt. Int. ∠ Thm. twice, followed by the ∠ Add. Post. $x° = 55° + 62° = 117°$.

4. It is possible to prove the Exterior Angle Theorem by drawing an auxiliary line. Use the figure at the right to show how this might be done. Write a complete proof on a separate sheet of paper. (Hint: Think about the angles formed when parallel lines are cut by a transversal.)

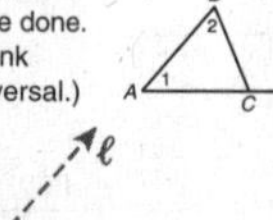

Additional Answer: Proofs will vary.

Given: △*ABC* with exterior angle ∠*BCD*

Prove: m∠*BCD* = m∠1 + m∠2

Proof:

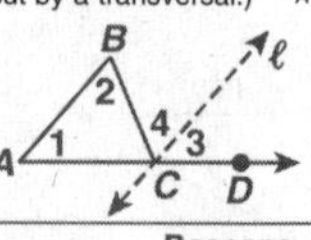

Statements	Reasons
1. ∠*ABC* with exterior angle ∠*BCD*	1. Given
2. Through *C*, draw line ℓ parallel to $\overline{AB}$.	2. Through a point outside a line, there is exactly one line parallel to the given line.
3. m∠1 = m∠3	3. ∥ lines → corr. ∠s ≅
4. m∠2 = m∠4	4. ∥ lines → alt. int. ∠s ≅
5. m∠*BCD* = m∠3 + m∠4	5. ∠ Add. Post.
6. m∠*BCD* = m∠1 + m∠2	6. Subst. Prop. of =

LESSON 4-2

Problem Solving

Angle Relationships in Triangles

1. The locations of three food stands on a fair's midway are shown. What is the measure of the angle labeled x°?

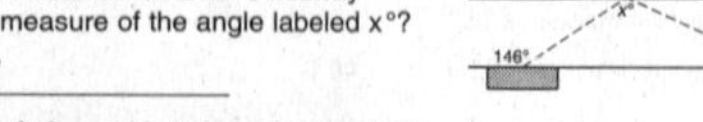

121°

2. A large triangular piece of plywood is to be painted to look like a mountain for the spring musical. The angles at the base of the plywood measure 76° and 45°. What is the measure of the top angle that represents the mountain peak?

59°

Use the figure of the banner for Exercises 3 and 4.

3. What is the value of n?

n = 12

4. What is the measure of each angle in the banner?

76°, 76°, 28°

Use the figure of the athlete pole vaulting for Exercises 5 and 6.

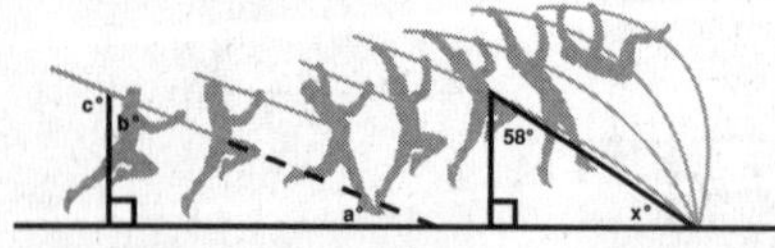

5. What is x°, the measure of the angle that the pole makes when it first touches the ground?

32°

6. At takeoff, a° = 23°. What is c°, the measure of the angle the pole makes with the athlete's body?

113°

The figure shows a path through a garden. Choose the best answer.

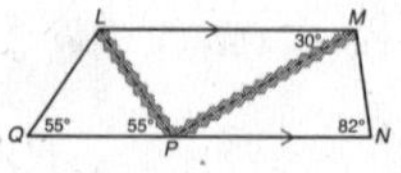

7. What is the measure of ∠QLP?

A 20° C 110°
(B) 70° D 125°

8. What is the measure of ∠LPM?

F 85° (H) 95°
G 90° J 125°

9. What is the measure of ∠PMN?

A 98° C 60°
(B) 68° D 55°

LESSON 4-2

Reading Strategies

Graphic Organizer

This graphic organizer describes the relationships of interior and exterior angles in a triangle.

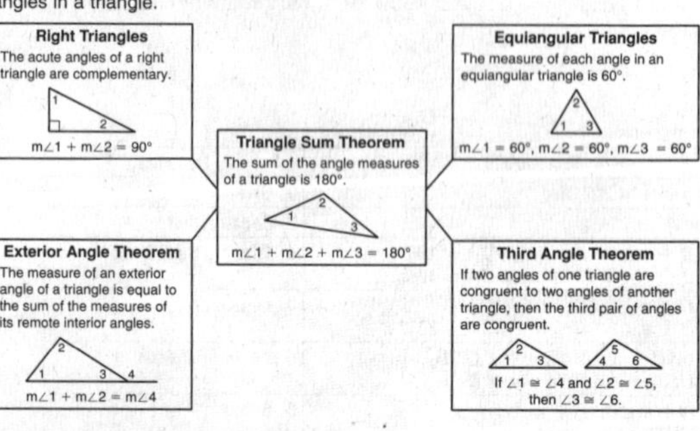

Use the given information to find the measures of the angles.

∠S and ∠Q are right angles.
m∠QPR = 30°
△TRP is equiangular.

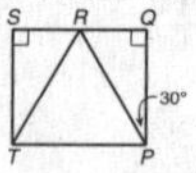

1. Find m∠QRP. 60°
2. Find m∠TRP. 60°
3. Find m∠RTS. 30°

Use the figure for Exercises 4–7.

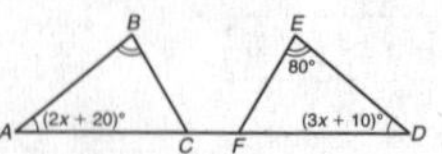

4. Find m∠A. 40°
5. Find m∠B. 80°
6. Find m∠BCF. 120°
7. Find m∠EFD. 60°

LESSON 4-3

Practice A

Congruent Triangles

Fill in the blanks to complete each definition.

1. **Corresponding** angles and **corresponding** sides are in the same position in polygons with an equal number of sides.
2. Two polygons are **congruent** polygons if and only if their corresponding angles and sides are congruent.

Refer to the figure of △GHI and △JKL for Exercises 3 and 4.

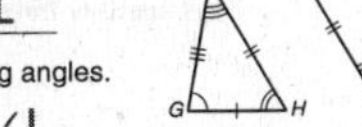

3. Name the three pairs of corresponding sides.
   $\overline{GH} \cong \overline{JK}$; $\overline{HI} \cong \overline{KL}$; $\overline{GI} \cong \overline{JL}$
4. Name the three pairs of corresponding angles.
   ∠G ≅ ∠J; ∠H ≅ ∠K; ∠I ≅ ∠L

Find the value of x.

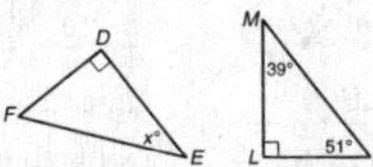

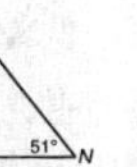

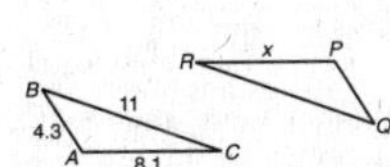

5. **Given:** △DEF ≅ △LMN **x = 39**
6. **Given:** △ABC ≅ △PQR **x = 8.1**
7. Etienne flies a kite. When the kite is flying well, the tail sticks out straight so the indicated angles at V are congruent. Use the phrases from the word bank to complete this two-column proof.

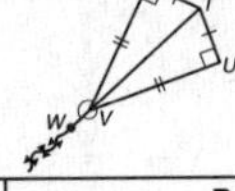

Given: ∠S and ∠U are right angles.
∠SVW ≅ ∠UVW, $\overline{SV} \cong \overline{UV}$, $\overline{ST} \cong \overline{UT}$

Prove: △STV ≅ △UTV

Word bank
∠S ≅ ∠U
Third ∠s Thm.
Given
∠SVT ≅ ∠UVT

Proof:

Statements	Reasons
1. $\overline{SV} \cong \overline{UV}$, $\overline{ST} \cong \overline{UT}$	1. Given
2. $\overline{TV} \cong \overline{TV}$	2. Reflex. Prop. of ≅
3. ∠S and ∠U are right angles.	3. a. **Given**
4. b. **∠S ≅ ∠U**	4. Rt. ∠ ≅ Thm.
5. ∠SVW ≅ ∠UVW	5. Given
6. ∠SVW and ∠SVT are supplementary, ∠UVW are ∠UVT are supplementary.	6. Lin. Pair Thm.
7. c. **∠SVT ≅ ∠UVT**	7. ≅ Suppls. Thm.
8. ∠STV ≅ ∠UTV	8. d. **Third ∠s Thm.**
9. △STV ≅ △UTV	9. Def. of ≅ △s

LESSON 4-3

Practice B

Congruent Triangles

In baseball, home plate is a pentagon. Pentagon ABCDE is a diagram of a regulation home plate. The baseball rules are very specific about the exact dimensions of this pentagon so that every home plate is congruent to every other home plate. If pentagon PQRST is another home plate, identify each congruent corresponding part.

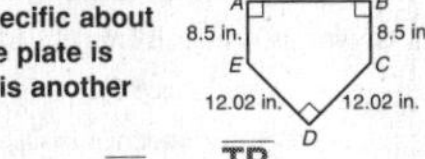

1. ∠S ≅ **∠D**
2. ∠B ≅ **∠Q**
3. $\overline{EA} \cong$ **$\overline{TP}$**
4. ∠E ≅ **∠T**
5. $\overline{PQ} \cong$ **$\overline{AB}$**
6. $\overline{TS} \cong$ **$\overline{ED}$**

Given: △DEF ≅ △LMN. Find each value.

7. m∠L = **40°**
8. EF = **37.3**
9. Write a two-column proof.

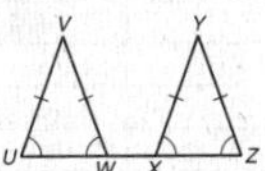

Given: ∠U ≅ ∠UWV ≅ ∠ZXY ≅ ∠Z, $\overline{UV} \cong \overline{WV}$, $\overline{XY} \cong \overline{ZY}$, $\overline{UX} \cong \overline{WZ}$

Prove: △UVW ≅ △XYZ

Proof: Possible answer:

Statements	Reasons
1. ∠U ≅ ∠UWV ≅ ∠ZXY ≅ ∠Z	1. Given
2. ∠V ≅ ∠Y	2. Third ∠s Thm.
3. $\overline{UV} \cong \overline{WV}$, $\overline{XY} \cong \overline{ZY}$	3. Given
4. $\overline{UX} \cong \overline{WZ}$	4. Given
5. UX = WZ, WX = WX	5. Def. of ≅ segs. Reflexive Prop. of =
6. UX = UW + WX, WZ = XZ + WX	6. Seg. Add. Post.
7. UW + WX = XZ + WX	7. Subst.
8. UW = XZ	8. Subtr. Prop. of =
9. △UVW ≅ △XYZ	9. Def. of ≅ △s

10. **Given:** △CDE ≅ △HIJ, DE = 9x, and IJ = 7x + 3. Find x and DE.
    $x = \frac{3}{2}$; $DE = 13\frac{1}{2}$
11. **Given:** △CDE ≅ △HIJ, m∠D = (5y + 1)°, and m∠I = (6y − 25)°. Find y and m∠D.
    y = 26; m∠D = 131°

LESSON 4-3

Practice C

Congruent Triangles

Mr. X is an inventive person. He takes pleasure in drawing a triangle and seeing if another person can recreate his drawing from piecemeal information. For each exercise, draw a diagram to support your answer. (Hint: Begin each exercise by drawing a triangle. Measure the parts of your triangle that Mr. X gives you and try to draw a different triangle with those parts. If the two triangles are congruent, you drew Mr. X's triangle.)

1. If Mr. X gives you the measures of the sides of a triangle, could you be sure you would draw Mr. X's triangle?
   Yes; possible answer:

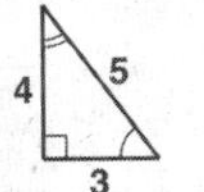

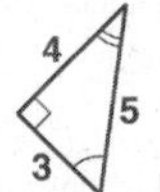

2. If Mr. X gives you the measures of the angles of a triangle, could you be sure you would draw Mr. X's triangle?
   No; possible answer:

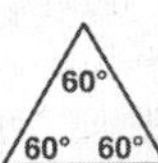

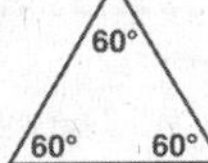

3. If Mr. X gives you the measures of one angle and of both sides of that angle, could you be sure you would draw Mr. X's triangle?
   Yes; possible answer:

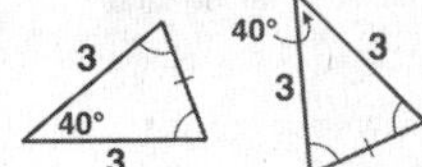

4. If Mr. X gives you the measures of one side and both angles that share that side, could you be sure you would draw Mr. X's triangle?
   Yes; possible answer:

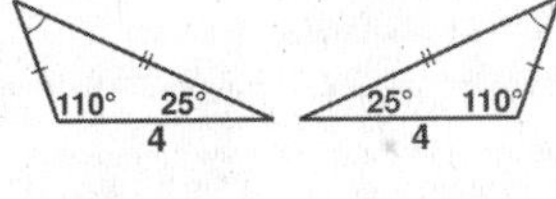

5. If Mr. X gives you the measures of one angle, one adjacent side, and the side opposite the angle, could you be sure you would draw Mr. X's triangle? (Hint: Start with an angle less than 45° and a long adjacent side.)
   No; possible answer:

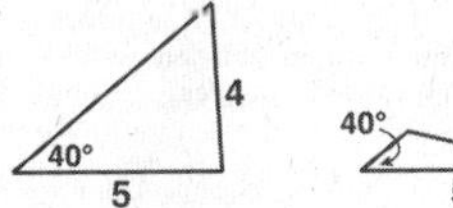

LESSON 4-3

Reteach

Congruent Triangles

Triangles are **congruent** if they have the same size and shape. Their **corresponding parts,** the angles and sides that are in the same positions, are congruent.

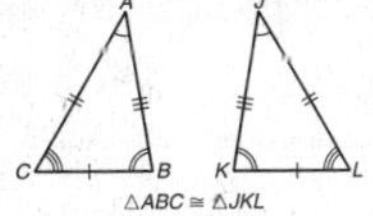

Corresponding Parts	
Congruent Angles	Congruent Sides
∠A ≅ ∠J	$\overline{AB} \cong \overline{JK}$
∠B ≅ ∠K	$\overline{BC} \cong \overline{KL}$
∠C ≅ ∠L	$\overline{CA} \cong \overline{LJ}$

To identify corresponding parts of congruent triangles, look at the order of the vertices in the congruence statement such as △ABC ≅ △JKL.

Given: △XYZ ≅ △NPQ. Identify the congruent corresponding parts.

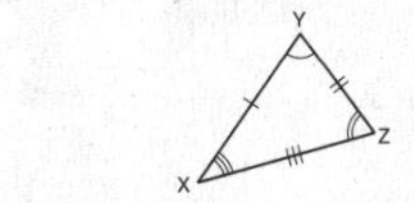

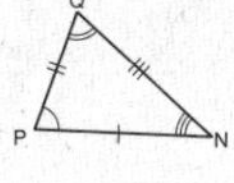

1. ∠Z ≅ **∠Q**
2. $\overline{YZ} \cong$ **$\overline{PQ}$**
3. ∠P ≅ **∠Y**
4. ∠X ≅ **∠N**
5. $\overline{NQ} \cong$ **$\overline{XZ}$**
6. $\overline{PN} \cong$ **$\overline{YX}$**

Given: △EFG ≅ △RST. Find each value below.

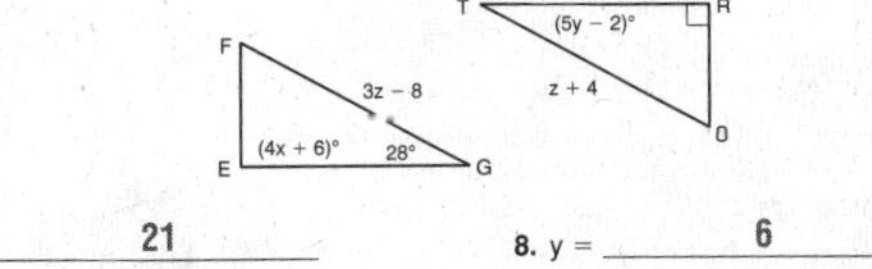

7. x = **21**
8. y = **6**
9. m∠F = **62°**
10. ST = **10**

LESSON 4-3

Reteach

Congruent Triangles continued

You can prove triangles congruent by using the definition of congruence.

Given: ∠*D* and ∠*B* are right angles.

∠*DCE* ≅ ∠*BCA*

C is the midpoint of $\overline{DB}$.

$\overline{ED} \cong \overline{AB}$, $\overline{EC} \cong \overline{AC}$

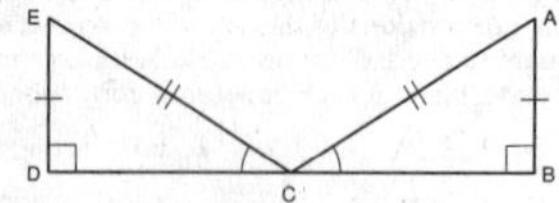

Prove: △*EDC* ≅ △*ABC*

Proof:

Statements	Reasons
1. ∠*D* and ∠*B* are rt. ∡.	1. Given
2. ∠*D* ≅ ∠*B*	2. Rt. ∠ ≅ Thm.
3. ∠*DCE* ≅ ∠*BCA*	3. Given
4. ∠*E* ≅ ∠*A*	4. Third ∡ Thm.
5. *C* is the midpoint of $\overline{DB}$.	5. Given
6. $\overline{DC} \cong \overline{BC}$	6. Def. of mdpt.
7. $\overline{ED} \cong \overline{AB}$, $\overline{EC} \cong \overline{AC}$	7. Given
8. △*EDC* ≅ △*ABC*	8. Def. of ≅ △s

11. Complete the proof.

Given: ∠*Q* ≅ ∠*R*

P is the midpoint of $\overline{QR}$.

$\overline{NQ} \cong \overline{SR}$, $\overline{NP} \cong \overline{SP}$

Prove: △*NPQ* ≅ △*SPR*

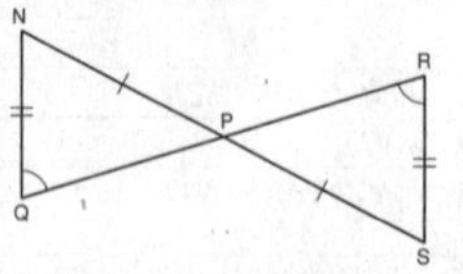

Proof:

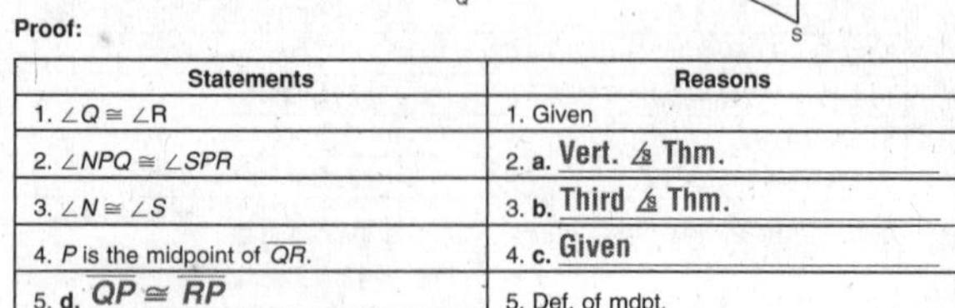

Statements	Reasons
1. ∠*Q* ≅ ∠R	1. Given
2. ∠*NPQ* ≅ ∠*SPR*	2. **a.** Vert. ∡ Thm.
3. ∠*N* ≅ ∠*S*	3. **b.** Third ∡ Thm.
4. *P* is the midpoint of $\overline{QR}$.	4. **c.** Given
5. **d.** $\overline{QP} \cong \overline{RP}$	5. Def. of mdpt.
6. $\overline{NQ} \cong \overline{SR}$, $\overline{NP} \cong \overline{SP}$	6. **e.** Given
7. △*NPQ* ≅ △*SPR*	7. **f.** Def. of ≅ △s

LESSON 4-3

Challenge

Congruence and Transformations on an Array

When two geometric figures are congruent, each is the image of the other under a rigid transformation. The first diagram at right shows two triangles, each positioned on an identical 3-by-3 array of dots. Are the triangles congruent?

In the second diagram at right, the triangles appear on the same array, and each dot is named as a point. The first triangle is △*BEG*, and the second is △*HEA*. It is clear that △*HEA* is a reflection of △*BEG* across a line, $\overleftrightarrow{DF}$. So △*HEA* is congruent to △*BEG*.

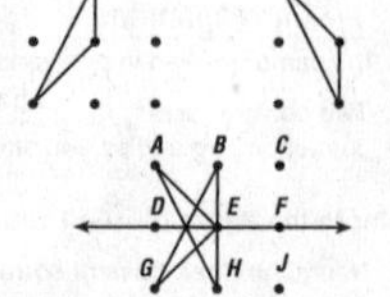

Each triangle is congruent to △*BEG*. Identify the transformation that relates the triangle to △*BEG*.

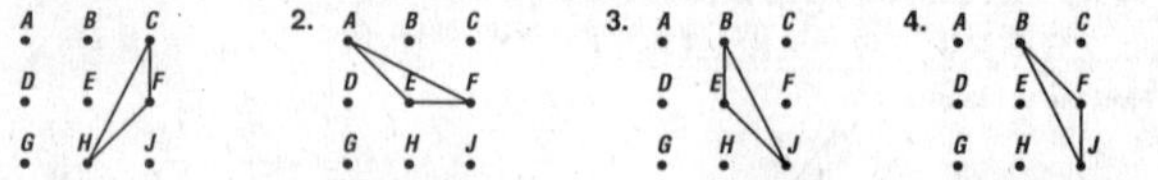

1. translation one unit right
2. rotation 90° clockwise about point *E*
3. reflection across $\overleftrightarrow{BH}$
4. glide reflection (reflection across $\overleftrightarrow{DF}$ and translation one unit right)

On each grid, sketch a triangle congruent to △*BEG* different from those given above. Use only the labeled points as vertices. Name the transformation that relates the new triangle to △*BEG*.

Figures will vary. Sample figures are given.

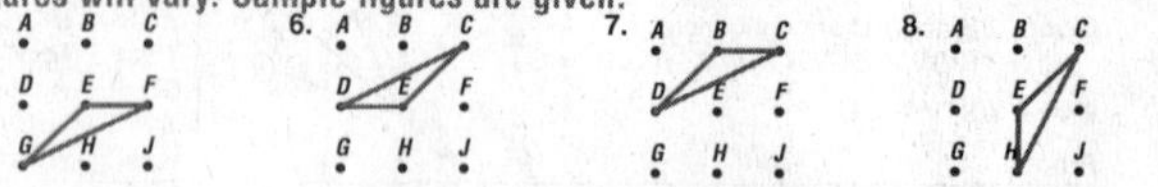

5. reflection across $\overleftrightarrow{CG}$
6. reflection across $\overleftrightarrow{AJ}$
7. reflection across $\overleftrightarrow{CG}$ and translation one unit up
8. rotation 180° about point *E*

9. Refer to the 3-by-3 grids in Exercises 1–8. Using the labeled points as vertices, how many triangles congruent to △*BEG* are there in all? List them.

   15: △*ADH*, △*GDB*, △*CFH*, △*JFB*, △*BEJ*, △*HEA*, △*HEC*, △*ABF*, △*CBD*, △*GHF*, △*JHD*, △*DEC*, △*DEJ*, △*FEA*, △*FEG*

10. Refer to the 3-by-3 grids in Exercises 1–8. Using the labeled points as vertices, how many triangles can be formed on each grid? List them on a separate sheet of paper, dividing the list into groups of congruent triangles.

    There are 76 triangles in all.

LESSON 4-3

Problem Solving

Congruent Triangles

Use the diagram of the fence for Exercises 1 and 2.

△*RQW* ≅ △*TVW*

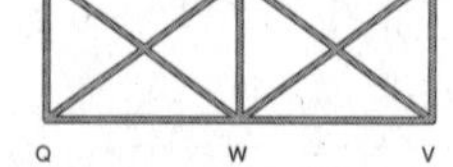

1. If m∠*RWQ* = 36° and m∠*TWV* = (2*x* + 5)°, what is the value of *x*?

   $x = 15.5$

2. If *RW* = (3*y* − 1) feet and *TW* = (*y* + 5) feet, what is the length of $\overline{RW}$?

   8 ft

Use the diagram of a section of the Bank of China Tower for Exercises 3 and 4.

△*JKL* ≅ △*LHJ*

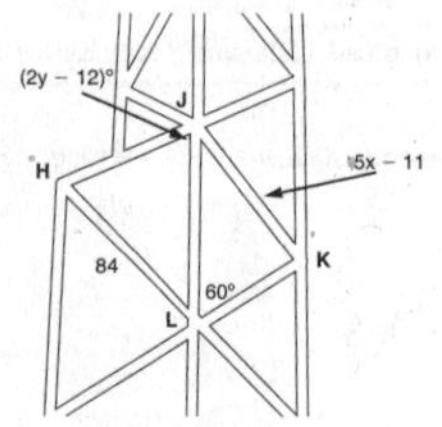

3. What is the value of *x*?

   $x = 19$

4. Find m∠*JHL*.

   72°

Choose the best answer.

5. Chairs with triangular seats were popular in the Middle Ages. Suppose a chair has a seat that is an isosceles triangle and the congruent sides measure $1\frac{1}{2}$ feet. A second chair has a triangular seat with a perimeter of $5\frac{1}{10}$ feet, and it is congruent to the first seat. What is a side length of the second seat?

   A $1\frac{4}{5}$ ft C 3 ft
   (B) $2\frac{1}{10}$ ft D $3\frac{3}{5}$ ft

Use the diagram for Exercises 6 and 7.

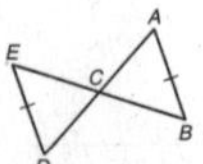

6. *C* is the midpoint of $\overline{EB}$ and $\overline{AD}$. What additional information would allow you to prove △*ABC* ≅ △*DEC* by the definition of congruent triangles?

   F $\overline{EB} \cong \overline{AD}$ H ∠*ECD* ≅ ∠*ACB*
   G $\overline{DE} \cong \overline{AB}$ **(J)** ∠*A* ≅ ∠*D*, ∠*B* ≅ ∠*E*

7. If △*ABC* ≅ △*DEC*, *ED* = 4*y* + 2, and *AB* = 6*y* − 4, what is the length of $\overline{AB}$?

   A 3 **(C)** 14
   B 12 D 18

LESSON 4-3

Reading Strategies

Understand Labels

Examine these two triangles.

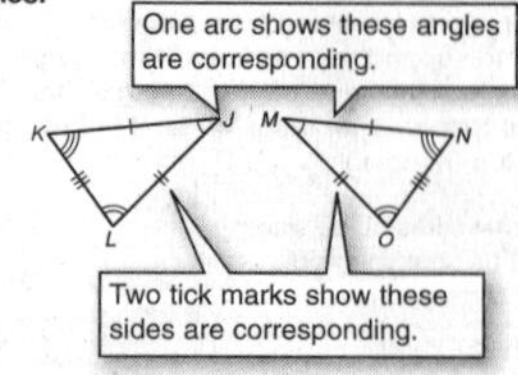

1. How can you tell which angle corresponds to ∠*L*?

   ∠*O* does because they both have two arcs.

2. How can you tell which side corresponds to $\overline{KL}$?

   It is side $\overline{NO}$ because both sides have three tick marks.

Answer the following questions based on these two triangles.

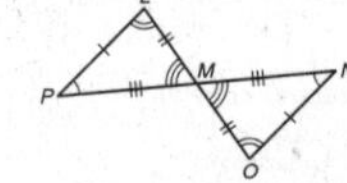

3. What angle corresponds to ∠*LMP*? **∠*OMN***
4. What angle corresponds to ∠*P*? **∠*N***
5. What side corresponds to $\overline{PL}$? **$\overline{NO}$**
6. What side corresponds to $\overline{LM}$? **$\overline{OM}$**

These two triangles are congruent. This statement can be written as follows: △*ABC* ≅ △*XYZ*.
Labeling triangles in this way is meaningful because it states that in these two triangles, ∠*A* ≅ ∠*X*; ∠*B* ≅ ∠*Y*; and ∠*C* ≅ ∠*Z*. The order in which the letters are placed tells which angles are congruent.

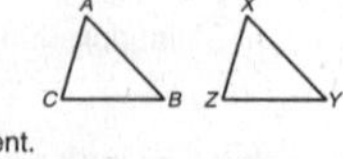

Answer the following questions based on these two triangles.

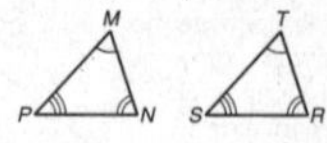

7. Write a congruence statement for these two triangles. **△*MNP* ≅ △*TRS***

8. How did you determine the order of the letters in your congruence statement?

   Corresponding angles of congruent triangles have the same measure, and the order of the letters indicates which angles are congruent.

LESSON 4-4

Practice A

Triangle Congruence: SSS and SAS

Name the included angle for each pair of sides.

1. $\overline{PQ}$ and $\overline{PR}$ **∠P**
2. $\overline{RQ}$ and $\overline{PR}$ **∠R**
3. $\overline{PQ}$ and $\overline{RQ}$ **∠Q**

Write SSS (Side-Side-Side Congruence) or SAS (Side-Angle-Side Congruence) next to the correct postulate.

4. If three sides of one triangle are congruent to three sides of another triangle, then the triangles are congruent. **SSS**
5. If two sides and the included angle of one triangle are congruent to two sides and the included angle of another triangle, then the triangles are congruent. **SAS**

GHIJ is a rectangle. A rectangle is a four-sided figure with four right angles and congruent opposite sides. For Exercises 6 and 7, fill in the blanks to show that △GHJ ≅ △IJH in two different ways.

6. It is given that **∠G** and **∠I** are right angles and that $\overline{GH}$ ≅ $\overline{JI}$ and $\overline{GJ}$ ≅ $\overline{HI}$. All right angles are congruent. So △GHJ ≅ △IJH by **SAS**.
7. It is given that $\overline{GH}$ ≅ **$\overline{JI}$** and $\overline{GJ}$ ≅ **$\overline{HI}$**. By the Reflexive Property of ≅, $\overline{JH}$ ≅ **$\overline{JH}$**. So △GHJ ≅ △IJH by **SSS**.
8. U.S. President Harry Truman and British Prime Minister Winston Churchill both wore polka-dot bow ties while in office. A well-tied bow tie resembles two congruent triangles. Use the phrases from the word bank to complete this two-column proof.

∠ABE ≅ ∠DBC,
SAS,
$\overline{BA}$ ≅ $\overline{BD}$,
$\overline{BE}$ ≅ $\overline{BC}$

Given: $\overline{BA}$ ≅ $\overline{BD}$, $\overline{BE}$ ≅ $\overline{BC}$

Prove: △ABE ≅ △DBC

Proof:

Statements	Reasons
1. a. **$\overline{BA}$ ≅ $\overline{BD}$, $\overline{BE}$ ≅ $\overline{BC}$**	1. Given
2. b. **∠ABE ≅ ∠DBC**	2. Vert. ∠s Thm.
3. △ABE ≅ △DBC	3. c. **SAS**

LESSON 4-4

Practice B

Triangle Congruence: SSS and SAS

Write which of the SSS or SAS postulates, if either, can be used to prove the triangles congruent. If no triangles can be proved congruent, write neither.

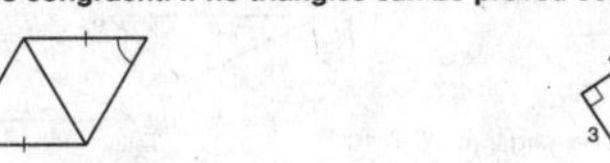

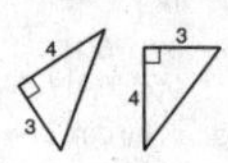

1. **neither**
2. **SAS**

3. **neither**
4. **SSS**

Find the value of x so that the triangles are congruent.

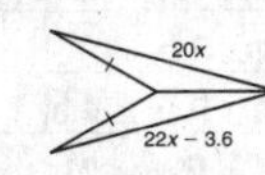

(6x − 27)° (4x + 7)°

5. x = **1.8**
6. x = **17**

The Hatfield and McCoy families are feuding over some land. Neither family will be satisfied unless the two triangular fields are exactly the same size. You know that C is the midpoint of each of the intersecting segments. Write a two-column proof that will settle the dispute.

7. **Given:** C is the midpoint of $\overline{AD}$ and $\overline{BE}$.

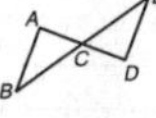

Prove: △ABC ≅ △DEC

Proof: **Possible answer:**

Statements	Reasons
1. C is the midpoint of $\overline{AD}$ and $\overline{BE}$.	1. Given
2. AC = CD, BC = CE	2. Def. of mdpt.
3. $\overline{AC}$ ≅ $\overline{CD}$, $\overline{BC}$ ≅ $\overline{CE}$	3. Def. of ≅ segs.
4. ∠ACB ≅ ∠DCE	4. Vert. ∠s Thm.
5. △ABC ≅ △DEC	5. SAS

LESSON 4-4

Practice C

Triangle Congruence: SSS and SAS

For each definition, tell what angle measures or side lengths of the quadrilateral must be given in order to determine a specific figure. (Hint: Think in terms of congruent triangles.)

1. A square is a quadrilateral with four congruent sides and four right angles.
   any side length
2. A rectangle is a quadrilateral with congruent pairs of opposite sides and four right angles.
   lengths of two adjacent sides
3. A rhombus is a quadrilateral with four congruent sides and parallel pairs of opposite sides.
   any angle measure and any side length
4. A parallelogram is a quadrilateral with congruent and parallel pairs of opposite sides.
   any angle measure and the lengths of two adjacent sides
5. Can a square be determined given only the length of a diagonal? Explain your answer.
   Yes; possible answer: The diagonal is the hypotenuse of an isosceles right triangle. The length of one side can be found by using the Pythagorean Theorem, and knowing one side is enough to draw a specific square.
6. Find the total area of the figure.
   (5x + 4) ft
   (x² − 10) ft
   15 ft
   540 ft²

Write a paragraph proof.

7. **Given:** $\overline{BA}$ ≅ $\overline{BC}$, $\overline{BE}$ ≅ $\overline{BF}$, $\overline{AG}$ ≅ $\overline{CD}$, $\overline{GF}$ ≅ $\overline{DE}$
   Prove: △AEG ≅ △CFD

Possible answer: It is given that $\overline{BA}$ ≅ $\overline{BC}$ and $\overline{BE}$ ≅ $\overline{BF}$, so by the definition of congruent segments, BA = BC and BE = BF. Adding these together gives BA + BE = BC + BF, and from the figure and the Segment Addition Postulate, AE = BA + BE and CF = BC + BF. It is clear by the Transitive Property that AE = CF, hence $\overline{AE}$ ≅ $\overline{CF}$ by the definition of ≅ segments. It is given that $\overline{GF}$ ≅ $\overline{DE}$ and the Reflexive Property shows that $\overline{FE}$ ≅ $\overline{FE}$. So by the Common Segments Theorem, $\overline{GE}$ ≅ $\overline{DF}$. The final pair of sides is given congruent, so △AEG ≅ △CFD by the Side-Side-Side Congruence Postulate.

LESSON 4-4

Reteach

Triangle Congruence: SSS and SAS

Side-Side-Side (SSS) Congruence Postulate

If three sides of one triangle are congruent to three sides of another triangle, then the triangles are congruent.

$\overline{QR}$ ≅ $\overline{TU}$, $\overline{RP}$ ≅ $\overline{US}$, and $\overline{PQ}$ ≅ $\overline{ST}$, so △*PQR* ≅ △*STU*.

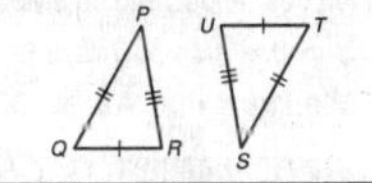

You can use SSS to explain why △*FJH* ≅ △*FGH*.
It is given that $\overline{FJ}$ ≅ $\overline{FG}$ and that $\overline{JH}$ ≅ $\overline{GH}$. By the Reflex. Prop. of ≅, $\overline{FH}$ ≅ $\overline{FH}$. So △*FJH* ≅ △*FGH* by SSS.

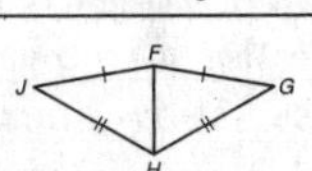

Side-Angle-Side (SAS) Congruence Postulate

If two sides and the included angle of one triangle are congruent to two sides and the included angle of another triangle, then the triangles are congruent.

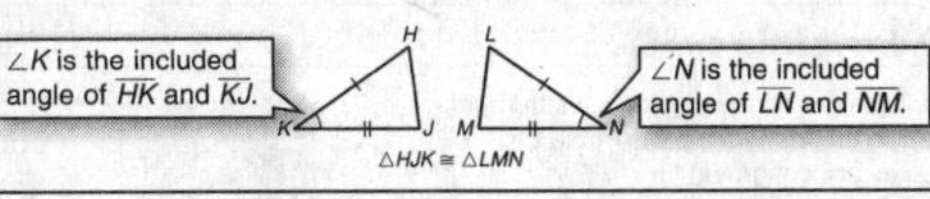

Use SSS to explain why the triangles in each pair are congruent.

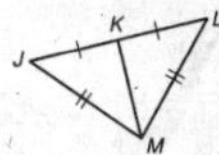

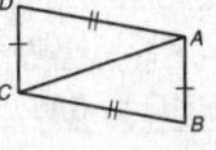

1. △*JKM* ≅ △*LKM*
   It is given that $\overline{JK}$ ≅ $\overline{LK}$ and that $\overline{JM}$ ≅ $\overline{LM}$. By the Reflex. Prop. of ≅, $\overline{KM}$ ≅ $\overline{KM}$. So △JKM ≅ △LKM by SSS.
2. △*ABC* ≅ △*CDA*
   It is given that $\overline{AB}$ ≅ $\overline{CD}$ and that $\overline{AD}$ ≅ $\overline{CB}$. By the Reflex. Prop. of ≅, $\overline{AC}$ ≅ $\overline{AC}$. So △ABC ≅ △CDA by SSS.
3. Use SAS to explain why △*WXY* ≅ △*WZY*.
   It is given that $\overline{ZW}$ ≅ $\overline{XW}$ and that ∠ZWY ≅ ∠XWY. By the Reflex. Prop. of ≅, $\overline{WY}$ ≅ $\overline{WY}$. So △WXY ≅ △WZY by SAS.

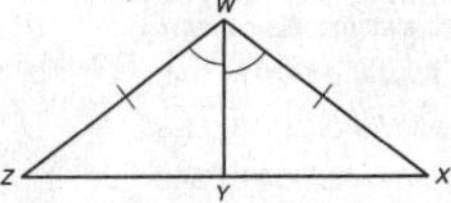

LESSON 4-4 **Reteach**

Triangle Congruence: SSS and SAS continued

You can show that two triangles are congruent by using SSS and SAS.

Show that $\triangle JKL \cong \triangle FGH$ for $y = 7$.

$HG = y + 6 = 7 + 6 = 13$ $m\angle G = 5y + 5 = 5(7) + 5 = 40°$ $FG = 4y - 1 = 4(7) - 1 = 27$

$HG = LK = 13$, so $\overline{HG} \cong \overline{LK}$ by def. of ≅ segs. $m\angle G = 40°$, so $\angle G \cong \angle K$ by def. of ≅ ∠s. $FG = JK = 27$, so $\overline{FG} \cong \overline{JK}$ by def. of ≅ segs. Therefore $\triangle JKL \cong \triangle FGH$ by SAS.

Show that the triangles are congruent for the given value of the variable.

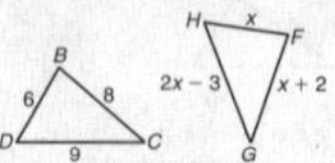

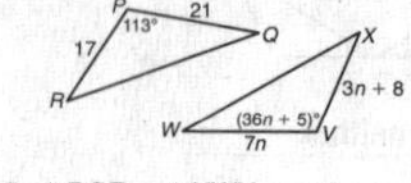

4. $\triangle BCD \cong \triangle FGH$, $x = 6$

BD = FH = 6, so $\overline{BD} \cong \overline{FH}$ by def. of ≅ segs. BC = FG = 8, so $\overline{BC} \cong \overline{FG}$ by def. of ≅ segs. CD = GH = 9, so $\overline{CD} \cong \overline{GH}$ by def. of ≅ segs. Therefore $\triangle BCD \cong \triangle FGH$ by SSS.

5. $\triangle PQR \cong \triangle VWX$, $n = 3$

PR = VX = 17, so $\overline{PR} \cong \overline{VX}$ by def. of ≅ segs. m∠P = m∠V = 113°, so ∠P ≅ ∠V by def. of ≅ ∠s. PQ = VW = 21, so $\overline{PQ} \cong \overline{VW}$ by def. of ≅ segs. Therefore $\triangle PQR \cong \triangle VWX$ by SAS.

6. Complete the proof.

Given: *T* is the midpoint of $\overline{VS}$. $\overline{RT} \perp \overline{VS}$

Prove: $\triangle RST \cong \triangle RVT$

Statements	Reasons
1. *T* is the midpoint of $\overline{VS}$.	1. Given
2. a. $\overline{VT} \cong \overline{ST}$	2. Def. of mdpt.
3. $\overline{RT} \perp \overline{VS}$	3. b. Given
4. ∠RTV and ∠RTS are rt. ∠s	4. c. Def. of ⊥ lines
5. d. ∠RTV ≅ ∠RTS	5. Rt. ∠ ≅ Thm.
6. $\overline{RT} \cong \overline{RT}$	6. e. Reflex. Prop. of ≅
7. $\triangle RST \cong \triangle RVT$	7. f. SAS

LESSON 4-4 **Challenge**

More Proofs with Overlapping Triangles

Write each two-column proof.

1. Given: $\overline{CE} \cong \overline{BE}$, $\overline{EA} \cong \overline{ED}$, $\overline{DC} \cong \overline{AB}$

Prove: $\triangle ABC \cong \triangle DCB$

Proof:

Statements	Reasons
1. $\overline{CE} \cong \overline{BE}$, $\overline{EA} \cong \overline{ED}$, $\overline{DC} \cong \overline{AB}$	1. Given
2. CE = BE, EA = ED, DC = AB	2. Def. of ≅ segments
3. CE + EA = BE + ED	3. Add. Prop. of =
4. CE + EA = CA, BE + ED = BD	4. Seg. Add. Post.
5. CA = BD	5. Subst.
6. $\overline{CA} \cong \overline{BD}$	6. Def. of ≅ segments
7. $\overline{CB} \cong \overline{CB}$	7. Reflex. Prop. of ≅
8. $\triangle ABC \cong \triangle DCB$	8. SSS (Steps 1, 6, 7)

2. Given: $\overline{LK} \cong \overline{HJ}$, $\overline{GK} \cong \overline{GJ}$, ∠GKL ≅ ∠GJH

Prove: $\triangle GLJ \cong \triangle GHK$

Proof:

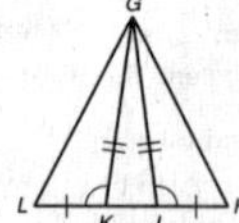

Statements	Reasons
1. $\overline{LK} \cong \overline{HJ}$, $\overline{GK} \cong \overline{GJ}$	1. Given
2. LK = HJ, GK = GJ	2. Def. of ≅ segments
3. KJ = KJ	3. Reflex. Prop. of =
4. LK + KJ = HJ + KJ	4. Add. Prop. of =
5. LK + KJ = LJ, HJ + KJ = HK	5. Seg. Add. Post.
6. LJ = HK	6. Subst.
7. $\overline{LJ} \cong \overline{HK}$	7. Def. of ≅ segments
8. ∠GKL ≅ ∠GJH	8. Given
9. ∠GKL and ∠GKJ are supplementary; ∠GJH and ∠GJK are supplementary.	9. Linear Pair Thm.
10. ∠GKJ ≅ ∠GJK	10. Congruent Supplements Thm.
11. $\triangle GLJ \cong \triangle GHK$	11. SAS (Steps 1, 7, 10)

LESSON 4-4 **Problem Solving**

Triangle Congruence: SSS and SAS

Use the diagram for Exercises 1 and 2.

A shed door appears to be divided into congruent right triangles.

1. Suppose $\overline{AB} \cong \overline{CD}$. Use SAS to show $\triangle ABD \cong \triangle DCA$.

We know that $\overline{AB} \cong \overline{DC}$. ∠ADC and ∠DAB are right angles, so ∠ADC ≅ ∠DAB by Rt. ∠ ≅ Thm. $\overline{AD} \cong \overline{DA}$ by Reflex. Prop. of ≅. So $\triangle ABD \cong \triangle DCA$ by SAS.

2. *J* is the midpoint of *AB* and $\overline{AK} \cong \overline{BK}$. Use SSS to explain why $\triangle AKJ \cong \triangle BKJ$.

We know that $\overline{AK} \cong \overline{BK}$. Since J is the midpoint of $\overline{AB}$, $\overline{AJ} \cong \overline{BJ}$ by def. of midpoint. $\overline{JK} \cong \overline{JK}$ by Reflex. Prop. of ≅. So $\triangle AKJ \cong \triangle BKJ$ by SSS.

3. A *balalaika* is a Russian stringed instrument. Show that the triangular parts of the two balalaikas are congruent for $x = 6$.

By the △ Sum Thm., m∠H = 54°. For x = 6, WY = FH = 10 in., m∠Y = m∠H = 54°, and XY = HG = 12 in. So $\triangle WXY \cong \triangle FHG$ by SAS.

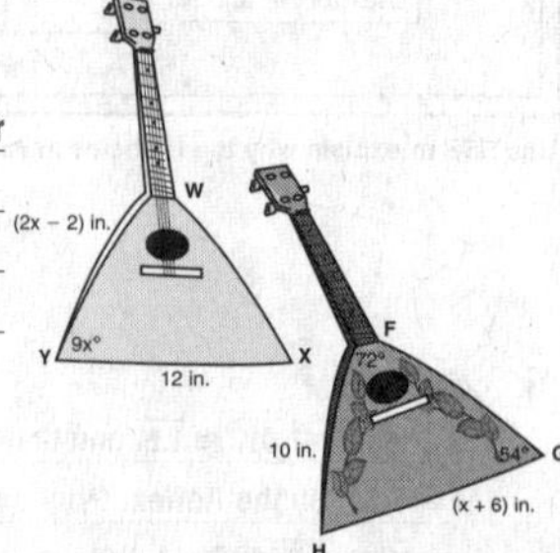

A quilt pattern of a dog is shown. Choose the best answer.

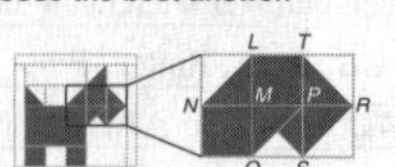

4. $ML = MP = MN = MQ = 1$ inch. Which statement is correct?

(A) $\triangle LMN \cong \triangle QMP$ by SAS.
B $\triangle LMN \cong \triangle QMP$ by SSS.
C $\triangle LMN \cong \triangle MQP$ by SAS.
D $\triangle LMN \cong \triangle MQP$ by SSS.

5. *P* is the midpoint of $\overline{TS}$ and $TR = SR = 1.4$ inches. What can you conclude about $\triangle TRP$ and $\triangle SRP$?

F $\triangle TRP \cong \triangle SRP$ by SAS.
(G) $\triangle TRP \cong \triangle SRP$ by SSS.
H $\triangle TRP \cong \triangle SPR$ by SAS.
J $\triangle TRP \cong \triangle SPR$ by SSS.

LESSON 4-4 **Reading Strategies**

Compare and Contrast

In mathematics, postulates and theorems are used to explain relationships. A **postulate** is a statement that is accepted without proof. A **theorem** is a statement that has been proven. Two postulates that can be used to prove that triangles are congruent are found in the following table:

Postulate	Hypothesis	Conclusion
SSS If three sides of one triangle are congruent to three sides of another triangle, then the triangles are congruent: Side-Side-Side Congruence	$\overline{XY} \cong \overline{QR}$, $\overline{XZ} \cong \overline{QS}$, $\overline{YZ} \cong \overline{RS}$	$\triangle XYZ \cong \triangle QRS$
SAS If two sides and the included angle of one triangle are congruent to two sides and the included angle of another triangle, then the triangles are congruent: Side-Angle-Side Congruence	$\overline{LM} \cong \overline{TU}$, $\overline{LN} \cong \overline{TV}$, ∠NLM ≅ ∠VTU	$\triangle LMN \cong \triangle TUV$

1. How is Postulate SSS like Postulate SAS?

Both involve the sides of the two triangles being compared.

2. How is Postulate SSS different from Postulate SAS?

Postulate SAS involves comparing included angles within the triangles, while SSS compares only the sides.

3. How is a postulate like a theorem?

Postulates and theorems are both statements that can be used to compare geometric shapes.

4. How are postulates and theorems different?

Postulates are accepted as being true without proof, while a theorem has been proven.

Determine whether each pair of triangles is congruent by SSS, SAS, or neither.

5. SSS

6. SAS

7. neither

8. SAS

LESSON 4-5

Practice A

Triangle Congruence: ASA, AAS, and HL

Name the included side for each pair of consecutive angles.

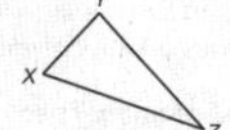

1. ∠X and ∠Z **$\overline{XZ}$**
2. ∠Y and ∠X **$\overline{YX}$**
3. ∠Y and ∠Z **$\overline{YZ}$**

Write *ASA* (Angle-Side-Angle Congruence), *AAS* (Angle-Angle-Side Congruence), or *HL* (Hypotenuse-Leg Congruence) next to the correct postulate.

4. If the hypotenuse and a leg of one right triangle are congruent to the hypotenuse and a leg of another right triangle, then the triangles are congruent. **HL**
5. If two angles and a nonincluded side of one triangle are congruent to the corresponding angles and nonincluded side of another triangle, then the triangles are congruent. **AAS**
6. If two angles and the included side of one triangle are congruent to two angles and the included side of another triangle, then the triangles are congruent. **ASA**

For Exercises 7–9, tell whether you can use each congruence theorem to prove that △ABC ≅ △DEF. If not, tell what else you need to know.

A, B, C, D, E, F

7. Hypotenuse-Leg
   No; you need to know that $\overline{AC} \cong \overline{DF}$.
8. Angle-Side-Angle
   Yes, if you use Third ∠s Thm. first.
9. Angle-Angle-Side
   Yes

10. A standard letter-sized envelope is a $9\frac{1}{2}$-in.-by-4-in. rectangle. The envelope is folded and glued from a sheet of paper shaped like the figure. Use the phrases in the word bank to complete this proof.

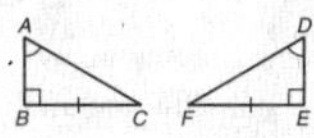

Given, ASA, Definition of rectangle

Given: *JMNK* is a rectangle. ∠IJK ≅ ∠LMN, ∠IKJ ≅ ∠LNM
Prove: △IJK ≅ △LMN

Statements	Reasons
1. ∠IJK ≅ ∠LMN, ∠IKJ ≅ ∠LNM	1. a. **Given**
2. $\overline{JK} \cong \overline{MN}$	2. b. **Definition of rectangle**
3. △IJK ≅ △LMN	3. c. **ASA**

LESSON 4-5

Practice B

Triangle Congruence: ASA, AAS, and HL

Students in Mrs. Marquez's class are watching a film on the uses of geometry in architecture. The film projector casts the image on a flat screen as shown in the figure. The dotted line is the bisector of ∠ABC. Tell whether you can use each congruence theorem to prove that △ABD ≅ △CBD. If not, tell what else you need to know.

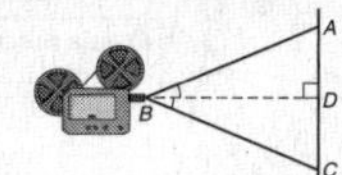

1. Hypotenuse-Leg
   No; you need to know that $\overline{AB} \cong \overline{CB}$.
2. Angle-Side-Angle
   Yes
3. Angle-Angle-Side
   Yes, if you use Third ∠s Thm. first.

Write which postulate, if any, can be used to prove the pair of triangles congruent.

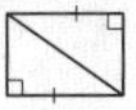

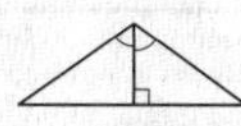

4. **HL**
5. **ASA or AAS**

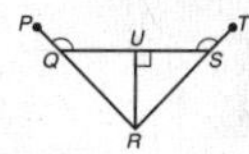

6. **none**
7. **AAS or ASA**

Write a paragraph proof.

8. **Given:** ∠PQU ≅ ∠TSU, ∠QUR and ∠SUR are right angles.
   Prove: △RUQ ≅ △RUS

P, U, T, Q, S, R

Possible answer: All right angles are congruent, so ∠QUR ≅ ∠SUR. ∠RQU and ∠PQU are supplementary and ∠RSU and ∠TSU are supplementary by the Linear Pair Theorem. But it is given that ∠PQU ≅ ∠TSU, so by the Congruent Supplements Theorem, ∠RQU ≅ ∠RSU. $\overline{RU} \cong \overline{RU}$ by the Reflexive Property of ≅, so △RUQ ≅ △RUS by AAS.

LESSON 4-5

Practice C

Triangle Congruence: ASA, AAS, and HL

1. In what case is knowing three parts (side lengths or angle measures) of a triangle not sufficient information to determine a specific triangle? **If the three parts are one non-right angle, one adjacent side, and the side opposite the angle, then you might not have enough information to draw the triangle; if the three parts are the three angles.**
2. Describe how the Leg-Leg (LL), Hypotenuse-Angle (HA), and Leg-Angle (LA) theorems of right-triangle congruence are related to the theorems and postulates of congruence that apply to all triangles. **Possible answer: In right triangles, one of the three angles is known, and all right angles are congruent. This means that fewer parts must be shown congruent to prove two right triangles congruent. LL is a special case of SAS in which the A is the right angle. HA is a special case of AAS in which the first A is the right angle. LA is a special case of ASA in which one A is a right angle.**
3. You know that a triangle is a right triangle. You are given two more parts. Is it possible that you do not have sufficient information to determine a specific triangle? Explain your answer.
   Yes; possible answer: If the two parts given are both angles, then the angle measures are set, but the side lengths are not.

Write a paragraph proof. Remember that if triangles are congruent, their matching parts are congruent.

4. **Given:** $\overline{GB}$, $\overline{GD}$, and $\overline{GF}$ are radii of the circle centered at *G* and are perpendicular to the sides of △ACE.
   Prove: △ACE is equilateral.

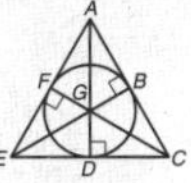

Possible answer: $\overline{GB}$, $\overline{GD}$, and $\overline{GF}$ all have the same length, by the definition of the radius of a circle, and so they are all congruent. Each radius is perpendicular to its side of the triangle, so all those angles are right angles, and they are all congruent. $\overline{GE}$ is congruent to $\overline{GE}$ by the Reflexive Property, and thus the conditions for HL congruence between △FGE and △DGE have been met. Similar reasoning shows that △FGA is congruent to △BGA and that △BGC is congruent to △DGC. ∠EGD and ∠AGB are vertical angles, so they are congruent. This fact, and the congruencies already shown, meet the conditions for ASA congruence between △EGD and △BGA. Similar reasoning shows that △FGE is congruent to △BGC and △FGA is congruent to △DGC. The Transitive Property of Congruence can now be used to show that all the triangles within △ACE are congruent. When triangles have been proven congruent, it is known that each matching part of the triangles is congruent. Hence AB = BC = CD = DE = EF = FA. By the Addition Property of Equality and the Segment Addition Postulate, AC = CE = EA and thus △ACE is equilateral.

LESSON 4-5

Reteach

Triangle Congruence: ASA, AAS, and HL

Angle-Side-Angle (ASA) Congruence Postulate

If two angles and the included side of one triangle are congruent to two angles and the included side of another triangle, then the triangles are congruent.

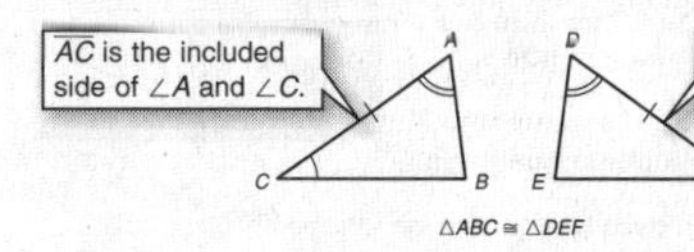

Determine whether you can use ASA to prove the triangles congruent. Explain.

L, N, 8 cm, P, 8 cm, K, M, Q

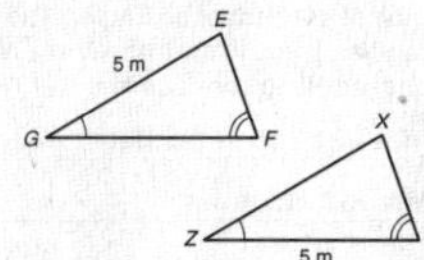

1. △KLM and △NPQ
   Yes; ∠K ≅ ∠N, $\overline{KL} \cong \overline{NP}$, and ∠L ≅ ∠P as given.
2. △EFG and △XYZ
   No; you need to know that $\overline{GF} \cong \overline{ZY}$.

P, K, N, M, L

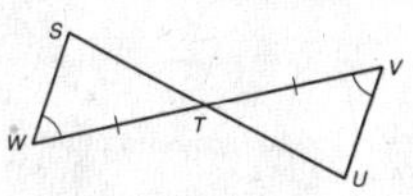

3. △KLM and △PNM, given that *M* is the midpoint of $\overline{NL}$
   No; you need to know that ∠NMP ≅ ∠LMK.
4. △STW and △UTV
   Yes; ∠W ≅ ∠V and $\overline{TW} \cong \overline{TV}$ as given. ∠STW ≅ ∠UTV by the Vert. ∠s Thm.

LESSON 4-5

Reteach

Triangle Congruence: ASA, AAS, and HL continued

Angle-Angle-Side (AAS) Congruence Theorem

If two angles and a nonincluded side of one triangle are congruent to the corresponding angles and nonincluded side of another triangle, then the triangles are congruent.

$\overline{FH}$ is a nonincluded side of $\angle F$ and $\angle G$.

$\overline{JL}$ is a nonincluded side of $\angle J$ and $\angle K$.

$\triangle FGH \cong \triangle JKL$

Special theorems can be used to prove right triangles congruent.

Hypotenuse-Leg (HL) Congruence Theorem

If the hypotenuse and a leg of a right triangle are congruent to the hypotenuse and a leg of another right triangle, then the triangles are congruent.

$\triangle JKL \cong \triangle MNP$

5. Describe the corresponding parts and the justifications for using them to prove the triangles congruent by AAS.

Given: $\overline{BD}$ is the angle bisector of $\angle ADC$.

Prove: $\triangle ABD \cong \triangle CBD$

$\angle A \cong \angle C$ (Given), $\angle ADB \cong \angle CDB$ (Def. of $\angle$ bisector),

$\overline{BD} \cong \overline{BD}$ (Reflex. Prop. of $\cong$)

Determine whether you can use the HL Congruence Theorem to prove the triangles congruent. If yes, explain. If not, tell what else you need to know.

6. $\triangle UVW \cong \triangle WXU$

Yes; $\overline{UV} \cong \overline{WX}$ (Given) and $\overline{UW} \cong \overline{UW}$ (Reflex. Prop. of $\cong$)

7. $\triangle TSR \cong \triangle PQR$

No; you need to know that $\overline{TR} \cong \overline{PR}$.

LESSON 4-5

Challenge

Congruent Triangles on the Coordinate Plane

When proving that two triangles are congruent, you may need to use your knowledge of lines on the coordinate plane.

Graph each set of lines. Then:
a. Identify two congruent triangles that are formed by the lines.
b. Write a paragraph proof to justify your statement in part a.

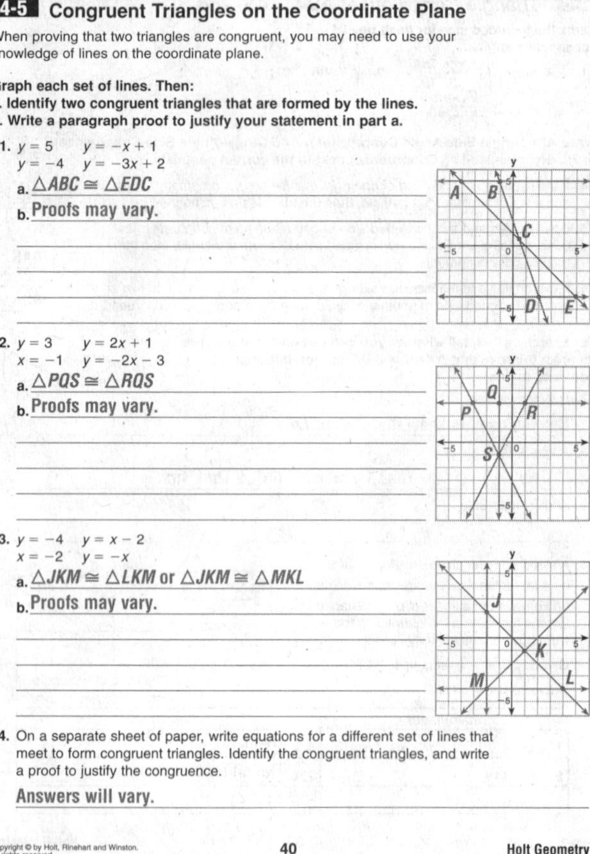

1. $y = 5$ $\quad y = -x + 1$
$y = -4$ $\quad y = -3x + 2$

a. $\triangle ABC \cong \triangle EDC$

b. Proofs may vary.

2. $y = 3$ $\quad y = 2x + 1$
$x = -1$ $\quad y = -2x - 3$

a. $\triangle PQS \cong \triangle RQS$

b. Proofs may vary.

3. $y = -4$ $\quad y = x - 2$
$x = -2$ $\quad y = -x$

a. $\triangle JKM \cong \triangle LKM$ or $\triangle JKM \cong \triangle MKL$

b. Proofs may vary.

4. On a separate sheet of paper, write equations for a different set of lines that meet to form congruent triangles. Identify the congruent triangles, and write a proof to justify the congruence.

Answers will vary.

LESSON 4-5

Problem Solving

Triangle Congruence: ASA, AAS, and HL

Use the following information for Exercises 1 and 2.

Melanie is at hole 6 on a miniature golf course. She walks east 7.5 meters to hole 7. She then faces south, turns 67° west, and walks to hole 8. From hole 8, she faces north, turns 35° west, and walks to hole 6.

Possible drawing: (6 —7.5 m— 7; angles 23°, 67°, 35°, 67°; 8)

1. Draw the section of the golf course described. Label the measures of the angles in the triangle.

2. Is there enough information given to determine the location of holes 6, 7, and 8? Explain.

Yes; the $\triangle$ is uniquely determined by AAS.

3. A section of the front of an English Tudor home is shown in the diagram. If you know that $\overline{KN} \cong \overline{LN}$ and $\overline{JN} \cong \overline{MN}$, can you use HL to conclude that $\triangle JKN \cong \triangle MLN$? Explain.

No; you need to know that $\angle KJN$ and $\angle LMN$ are rt. $\triangle$s.

Use the diagram of a kite for Exercises 4 and 5.

$\overline{AE}$ is the angle bisector of $\angle DAF$ and $\angle DEF$.

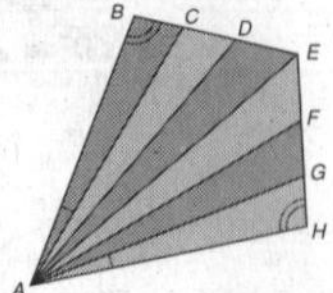

4. What can you conclude about $\triangle DEA$ and $\triangle FEA$?

A $\triangle DEA \cong \triangle FEA$ by HL.
B $\triangle DEA \cong \triangle FEA$ by AAA.
(C) $\triangle DEA \cong \triangle FEA$ by ASA.
D $\triangle DEA \cong \triangle FEA$ by SAS.

5. Based on the diagram, what can you conclude about $\triangle BCA$ and $\triangle HGA$?

F $\triangle BCA \cong \triangle HGA$ by HL.
G $\triangle BCA \cong \triangle HGA$ by AAS.
H $\triangle BCA \cong \triangle HGA$ by ASA.
(J) It cannot be shown using the given information that $\triangle BCA \cong \triangle HGA$.

LESSON 4-5

Reading Strategies

Use a Graphic Aid

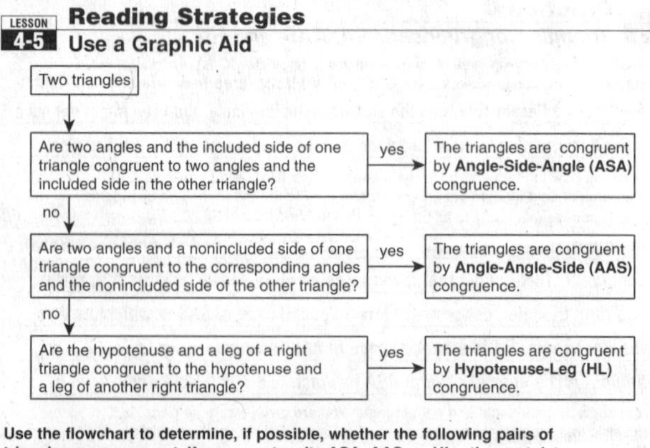

Use the flowchart to determine, if possible, whether the following pairs of triangles are congruent. If congruent, write ASA, AAS, or HL—the postulate you used to conclude that they are congruent. If it is not possible to conclude that they are congruent, write no conclusion.

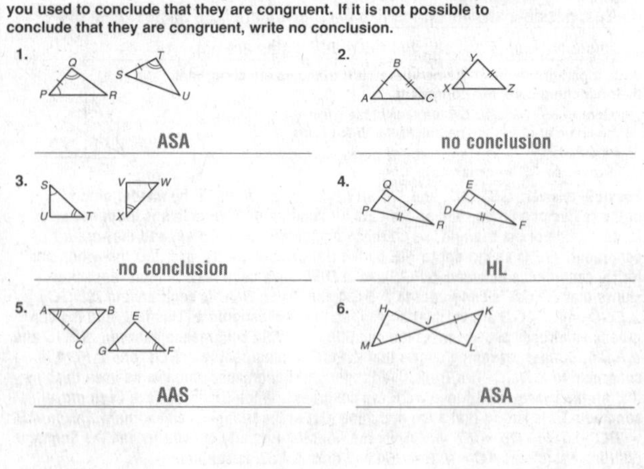

1. ASA

2. no conclusion

3. no conclusion

4. HL

5. AAS

6. ASA

LESSON 4-6

Practice A

Triangle Congruence: CPCTC

1. CPCTC is an abbreviation of the phrase "Corresponding **Parts** of Congruent **Triangles** are Congruent."

Use the figure for Exercises 2 and 3.
∠B ≅ ∠E, ∠C ≅ ∠F, and $\overline{AB} \cong \overline{DE}$.

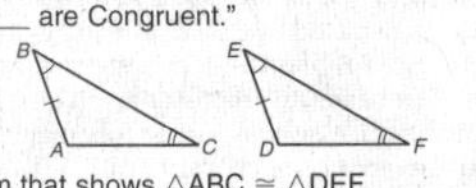

2. Name the triangle congruence theorem that shows △ABC ≅ △DEF. **AAS**

3. Use CPCTC to name the other three pairs of congruent parts in the triangles.

∠A ≅ **∠D**   **$\overline{AC}$** ≅ **$\overline{DF}$**   **$\overline{BC}$** ≅ **$\overline{EF}$**

4. Some hikers come to a river in the woods. They want to cross the river but decide to find out how wide it is first. So they set up congruent right triangles. The figure shows the river and the triangles. Find the width of the river, GH. **5 m**

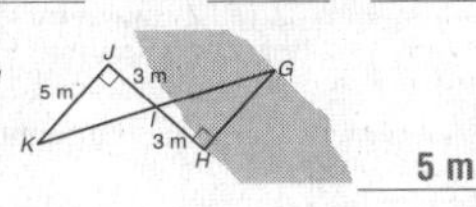

Use the phrases in the word bank to complete this proof.

5. Given: $\overline{PQ} \cong \overline{RQ}$, ∠PQS ≅ ∠RQS
Prove: ∠P ≅ ∠R

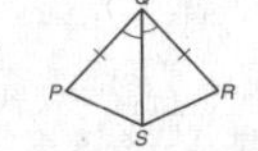

Word bank: $\overline{QS} \cong \overline{QS}$, CPCTC, △PQS ≅ △RQS

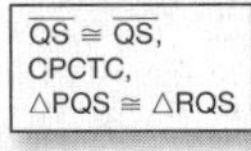

Statements	Reasons
1. $\overline{PQ} \cong \overline{RQ}$, ∠PQS ≅ ∠RQS	1. Given
2. a. **$\overline{QS} \cong \overline{QS}$**	2. Reflexive Property of ≅
3. b. **△PQS ≅ △RQS**	3. SAS
4. ∠P ≅ ∠R	4. c. **CPCTC**

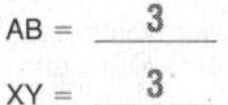

Use the blank graph for Exercises 6–9.

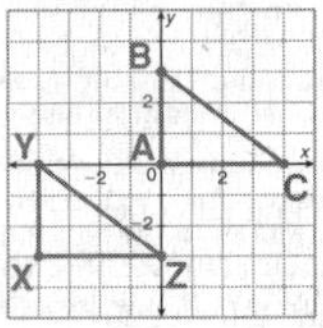

6. Plot these points: A(0, 0), B(0, 3), C(4, 0), X(−4, −3), Y(−4, 0), Z(0, −3). Draw segments to make △ABC and △XYZ.

7. Use the Distance Formula to find the length of each side.

AB = **3**   BC = **5**   AC = **4**

XY = **3**   YZ = **5**   XZ = **4**

8. Name the triangle congruence theorem that shows △ABC ≅ △XYZ. **SSS**

9. How do you know that ∠X ≅ ∠A? **CPCTC**

LESSON 4-6

Practice B

Triangle Congruence: CPCTC

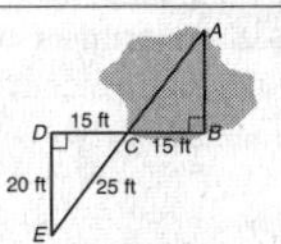

1. Heike Dreschler set the Woman's World Junior Record for the long jump in 1983. She jumped about 23.4 feet. The diagram shows two triangles and a pond. Explain whether Heike could have jumped the pond along path BA or along path CA. **Possible answer: Because ∠DCE ≅ ∠BCA by the Vertical ∡ Thm. the triangles are congruent by ASA, and each side in △ABC has the same length as its corresponding side in △EDC. Heike could jump about 23 ft. The distance along path BA is 20 ft because *BA* corresponds with *DE*, so Heike could have jumped this distance. The distance along path CA is 25 ft because *CA* corresponds with *CE*, so Heike could not have jumped this distance.**

Write a flowchart proof.

2. Given: ∠L ≅ ∠J, $\overline{KJ} \parallel \overline{LM}$
Prove: ∠LKM ≅ ∠JMK

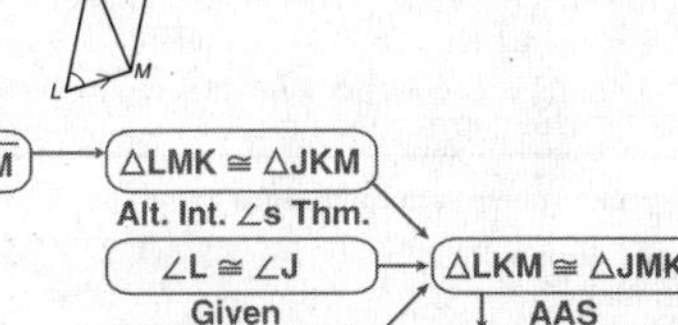

Write a two-column proof.

3. Given: FGHI is a rectangle.
Prove: The diagonals of a rectangle have equal lengths. **Possible answer:**

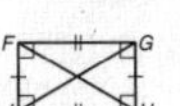

Statements	Reasons
1. FGHI is a rectangle.	1. Given
2. FI ≅ GH, ∠FIH and ∠GHI are right angles.	2. Def. of rectangle
3. ∠FIH ≅ ∠GHI	3. Rt. ∠ ≅ Thm.
4. IH ≅ IH	4. Reflex. Prop. of ≅
5. △FIH ≅ △GHI	5. SAS
6. FH ≅ GI	6. CPCTC
7. FH = GI	7. Def. of ≅ segs.

LESSON 4-6

Practice C

Triangle Congruence: CPCTC

Write paragraph proofs for Exercises 1 and 2.

1. Given: ABCD is a parallelogram.
Prove: The diagonals of a parallelogram bisect each other.

Possible answer: From the definition of a parallelogram, $\overline{DC}$ is congruent to $\overline{AB}$ and $\overline{DC}$ is parallel to $\overline{AB}$. By the Alternate Interior Angles Theorem, ∠BAC is congruent to ∠DCA and ∠CDB is congruent to ∠ABD. Therefore △ABE is congruent to △CDE by ASA. By CPCTC, $\overline{DE}$ is congruent to $\overline{BE}$ and $\overline{AE}$ is congruent to $\overline{CE}$. Congruent segments have equal lengths, so the diagonals bisect each other.

2. Given: FGHI is a rhombus.
Prove: The diagonals of a rhombus are congruent, perpendicular, and bisect the vertex angles of the rhombus.

(Note: Be careful naming the triangles. The order of vertices matters.)
Possible answer: From the definition of a rhombus, $\overline{IH}$ is congruent to $\overline{FG}$, *IF* is congruent to *GH*, and $\overline{IH}$ is parallel to $\overline{FG}$. By Alternate Interior Angles Theorem, ∠GFH is congruent to ∠IHF and ∠FGI is congruent to ∠HIG. Therefore △FGJ is congruent to △HIJ by ASA. By CPCTC, $\overline{FJ}$ is congruent to $\overline{HJ}$ and $\overline{GJ}$ is congruent to $\overline{IJ}$. So △FJI is congruent to △GHJ by SSS. But △HIJ is also congruent to △FIJ by SSS. And so all four triangles are congruent by the Transitive Property of Congruence. By CPCTC and the Segment Addition Postulate, $\overline{FH}$ is congruent to $\overline{GI}$. By CPCTC and the Linear Pair Theorem, ∠FJI, ∠GJF, ∠HJG, and ∠IJH are right angles. So $\overline{FH}$ and $\overline{GI}$ are perpendicular. By CPCTC, ∠GFH, ∠IFH, ∠GHF, and ∠IHF are congruent, so $\overline{FH}$ bisects ∠IFG and ∠IHG. Similar reasoning shows that $\overline{GI}$ bisects ∠FGH and ∠FIH.

3. Rectangles, rhombuses, and squares are all types of parallelograms. Write a conjecture about the diagonals of a rectangle.
The diagonals of a rectangle bisect each other.

4. A square is a type of rhombus. Write a conjecture about the diagonals of a square.
The diagonals of a square are congruent perpendicular bisectors that bisect the vertex angles of the square.

5. An isosceles trapezoid has one pair of noncongruent parallel sides, a pair of congruent nonparallel sides, and two pairs of congruent angles. What relationship do the diagonals of an isosceles trapezoid have?
The diagonals are congruent.

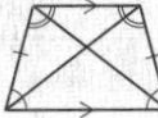

LESSON 4-6

Reteach

Triangle Congruence: CPCTC

Corresponding Parts of Congruent Triangles are Congruent **(CPCTC)** is useful in proofs. If you prove that two triangles are congruent, then you can use CPCTC as a justification for proving corresponding parts congruent.

Given: $\overline{AD} \cong \overline{CD}$, $\overline{AB} \cong \overline{CB}$
Prove: ∠A ≅ ∠C
Proof:

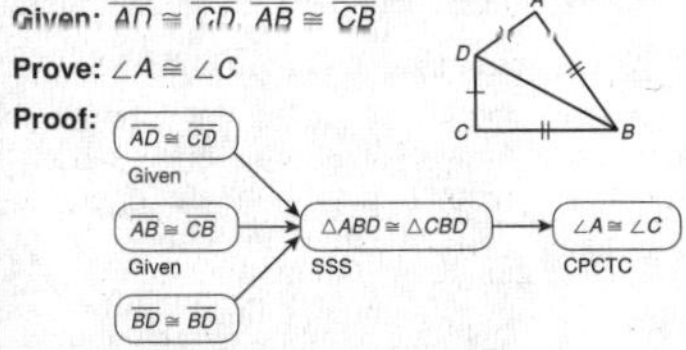

Complete each proof.

1. Given: ∠PNQ ≅ ∠LNM, $\overline{PN} \cong \overline{LN}$, *N* is the midpoint of $\overline{QM}$.
Prove: $\overline{PQ} \cong \overline{LM}$
Proof:

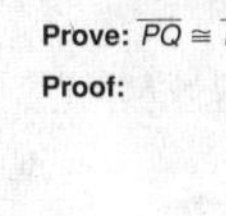

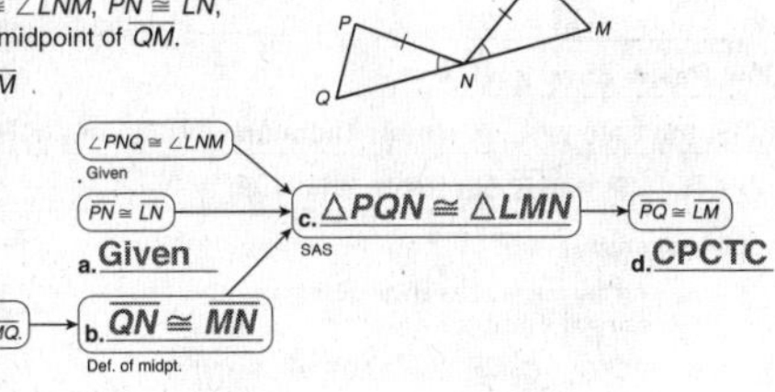

2. Given: △UXW and △UVW are right △s. $\overline{UX} \cong \overline{UV}$
Prove: ∠X ≅ ∠V 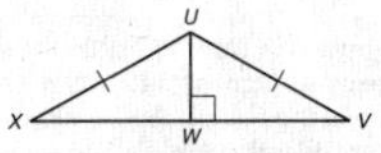
Proof:

Statements	Reasons
1. △UXW and △UVW are rt. △s.	1. Given
2. $\overline{UX} \cong \overline{UV}$	2. a. **Given**
3. $\overline{UW} \cong \overline{UW}$	3. b. **Reflex. Prop. of ≅**
4. c. **△UXW ≅ △UVW**	4. d. **HL**
5. ∠X ≅ ∠V	5. e. **CPCTC**

LESSON 4-6
Reteach
Triangle Congruence: CPCTC continued

You can also use CPCTC when triangles are on the coordinate plane.

Given: $C(2, 2)$, $D(4, -2)$, $E(0, -2)$, $F(0, 1)$, $G(-4, -1)$, $H(-4, 3)$

Prove: $\angle CED \cong \angle FHG$

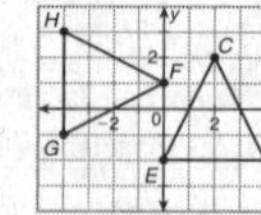

Step 1 Plot the points on a coordinate plane.

Step 2 Find the lengths of the sides of each triangle. Use the Distance Formula if necessary.

$d = \sqrt{(x_2 - x_1)^2 + (y_2 - y_1)^2}$

$CD = \sqrt{(4-2)^2 + (-2-2)^2} = \sqrt{4+16} = 2\sqrt{5}$ $\quad FG = \sqrt{(-4-0)^2 + (-1-1)^2} = \sqrt{16+4} = 2\sqrt{5}$

$DE = 4$ $\quad GH = 4$

$EC = \sqrt{(2-0)^2 + [2-(-2)]^2} = \sqrt{4+16} = 2\sqrt{5}$ $\quad HF = \sqrt{[0-(-4)]^2 + (1-3)^2} = \sqrt{16+4} = 2\sqrt{5}$

So, $\overline{CD} \cong \overline{FG}$, $\overline{DE} \cong \overline{GH}$, and $\overline{EC} \cong \overline{HF}$. Therefore $\triangle CDE \cong \triangle FGH$ by SSS, and $\angle CED \cong \angle FHG$ by CPCTC.

Use the graph to prove each congruence statement.

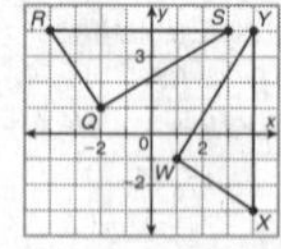

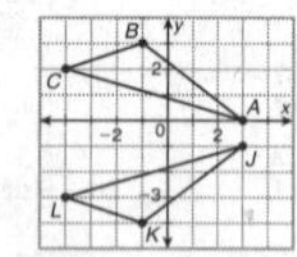

3. $\angle RSQ \cong \angle XYW$

$QR = WX = \sqrt{13}$, $RS = XY = 7$, $SQ = YW = \sqrt{34}$. So $\triangle QRS \cong \triangle WXY$ by SSS, and $\angle RSQ \cong \angle XYW$ by CPCTC.

4. $\angle CAB \cong \angle LJK$

$AB = JK = 5$, $BC = KL = \sqrt{10}$, $CA = LJ = \sqrt{53}$. So $\triangle ABC \cong \triangle JKL$ by SSS, and $\angle CAB \cong \angle LJK$ by CPCTC.

5. Use the given set of points to prove $\angle PMN \cong \angle VTU$.

$M(-2, 4)$, $N(1, -2)$, $P(-3, -4)$, $T(-4, 1)$, $U(2, 4)$, $V(4, 0)$

$MN = TU = 3\sqrt{5}$, $NP = UV = 2\sqrt{5}$, $PM = VT = \sqrt{65}$. So $\triangle MNP \cong \triangle TUV$ by SSS, and $\angle PMN \cong \angle VTU$ by CPCTC.

LESSON 4-6
Challenge
Isosceles Triangles and Roof Trusses

The wooden or metal framework that supports a roof is called a roof truss. The simplest type of roof truss has the shape of an isosceles triangle, as depicted by $\triangle ABC$ at right. In the diagram, the legs of the triangle, $\overline{AB}$ and $\overline{CB}$, represent sloping beams that are called rafters. The base, $\overline{AC}$, represents the tie beam that "ties together" the rafters. However, large roofs require trusses with designs that are more complex than this.

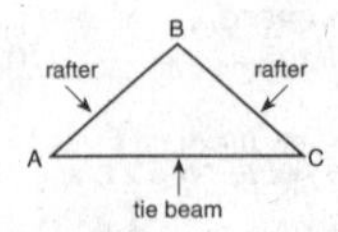

For example, a king-post truss is pictured at right. The king post, $\overline{KZ}$, is a median of $\triangle JKL$, and it provides support for the rafters. Additional support for the rafters comes from struts $\overline{ZX}$ and $\overline{ZY}$, which are medians of $\triangle JKZ$ and $\triangle LKZ$, respectively. The outer triangle, $\triangle JKL$, is an isosceles triangle with base $\overline{JL}$.

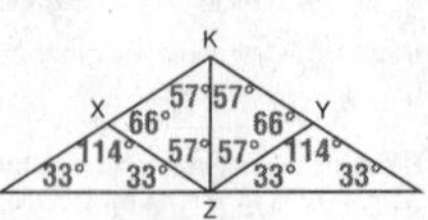

1. Refer to the diagram of the king-post truss. Write a flowchart proof to show that $\angle JKZ \cong \angle LKZ$.

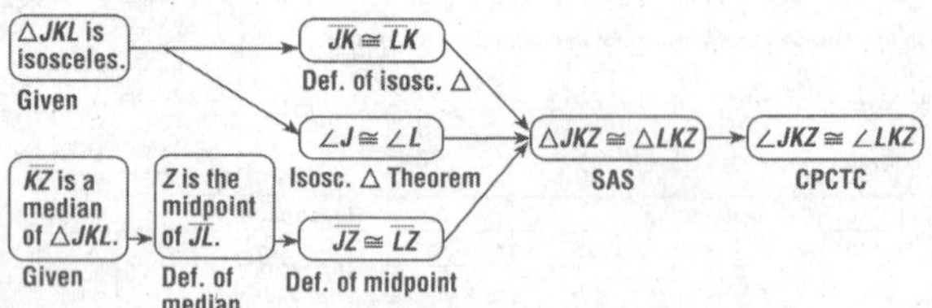

2. Explain why it must be true that $\angle XJZ \cong \angle XZJ$. (Hint: How is $\overline{ZX}$ related to $\triangle JKL$?)

Explanations may vary.

3. Given that $m\angle KJZ = 33°$, find the measures of all the other angles between the rafters, beams, and struts of the king-post truss. Label the angle measures directly on the figure.

In the queen-post truss pictured at right, congruent queen posts $\overline{NV}$ and $\overline{TW}$ are positioned so that they are perpendicular to tie beam $\overline{PR}$ and so that $\overline{PV} \cong \overline{VW} \cong \overline{WR}$. Struts $\overline{MV}$, $\overline{SW}$, and $\overline{NT}$ intersect the rafters so that $\overline{PM} \cong \overline{MN} \cong \overline{NQ}$ and $\overline{RS} \cong \overline{ST} \cong \overline{TQ}$. The outer triangle, $\triangle PQR$, is an isosceles triangle with base $\overline{PR}$.

4. Using the information about the queen-post truss given above, prove each statement on a separate sheet of paper. Use any form of proof that you want. **Proofs will vary.**

a. $\triangle QNT$ is isosceles. b. $\triangle MPV$ is isosceles. c. $\triangle VMN$ is isosceles.

5. Given that $m\angle NQT = 110°$, find the measures of all the other angles between the rafters, beams, and struts of the queen-post truss. Label the angle measures directly on the figure.

LESSON 4-6
Problem Solving
Triangle Congruence: CPCTC

1. Two triangular plates are congruent. The area of one of the plates is 60 square inches. What is the area of the other plate? Explain.

60 in^2; Since the triangles are ≅, they have the same measures. So, the triangles also have the same areas.

2. An archaeologist draws the triangles to find the distance XY across a ravine. What is XY? Explain.

82 m; $\triangle UVW \cong \triangle XYW$ by SAS, so $\overline{UV} \cong \overline{XY}$ by CPCTC. Therefore $UV = XY = 82$ m.

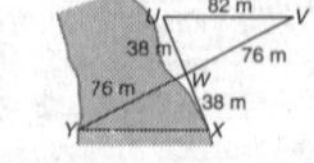

3. A city planner sets up the triangles to find the distance RS across a river. Describe the steps that she can use to find RS.

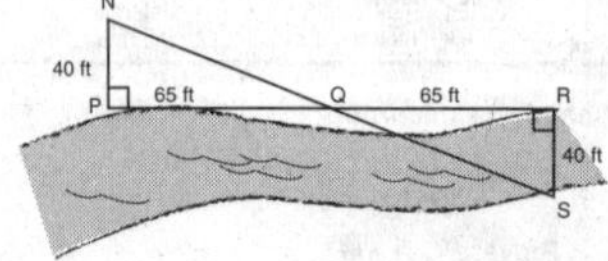

$\angle P \cong \angle R$ because they are both rt. ∡. $\overline{PQ} \cong \overline{RQ}$ because $PQ = RQ = 65$ ft. $\angle NQP \cong \angle SQR$ because vert. ∡ are ≅. Therefore $\triangle NPQ \cong \triangle SRQ$ by ASA. By CPCTC, $\overline{NP} \cong \overline{SR}$. So $SR = NP = 40$ ft.

Choose the best answer.

4. A lighthouse and the range of its shining light are shown. What can you conclude?

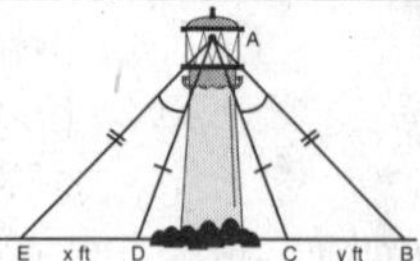

(A) x = y by CPCTC
C $\angle AED \cong \angle ADE$ by CPCTC
B x = 2y
D $\angle AED \cong \angle ACB$

5. A rectangular piece of cloth 15 centimeters long is cut along a diagonal to form two triangles. One of the triangles has a side length of 9 centimeters. Which is a true statement?

F The second triangle has an angle measure of 15° by CPCTC.
(G) The second triangle has a side length of 9 centimeters by CPCTC.
H You cannot make a conclusion about the side length of the second triangle.
J The triangles are not congruent.

6. Small sandwiches are cut in the shape of right triangles. The longest sides of all the sandwiches are 3 inches. One sandwich has a side length of 2 inches. Which is a true statement?

A All the sandwiches have a side length of 2 inches by CPCTC.
B All the sandwiches are isosceles triangles with side lengths of 2 inches.
C None of the other sandwiches have side lengths of 2 inches.
(D) You cannot make a conclusion using CPCTC.

LESSON 4-6
Reading Strategies
Using an Acronym

An **acronym** is a word formed from the first letters of a phrase. For example, ASAP stands for "As Soon As Possible." Acronyms can also combine the first letters or series of letters in a series of words, as in radar, which stands for **ra**dio **d**etecting **a**nd **r**anging.

One acronym used in geometry is CPCTC. Look at the breakdown of this acronym:

C P C T C

Corresponding
Parts of
Congruent
Triangles are
Congruent

1. What are some reasons you would use an acronym?

Possible answer: to abbreviate or make a statement simpler to understand or remember

2. What are some other acronyms you have used in your everyday life?

Answers will vary. Students may mention FBI, IRS, RSVP, FAQ, or others that are popular in text messaging.

Examine the figure and answer the question.

3. In this triangle, $\angle C \cong \angle N$ and $\overline{AC} \cong \overline{LN}$. Assume that $\triangle ABC \cong \triangle LMN$. Name four other parts that are congruent using CPCTC.

$\angle B \cong \angle M$; $\angle A \cong \angle L$; $\overline{CB} \cong \overline{NM}$; $\overline{AB} \cong \overline{LM}$

LESSON 4-7

Practice A

Introduction to Coordinate Proof

Draw a square with a side length of 4 units in the coordinate plane. Label the coordinates of each vertex.

1. Center one side at the origin (0, 0).

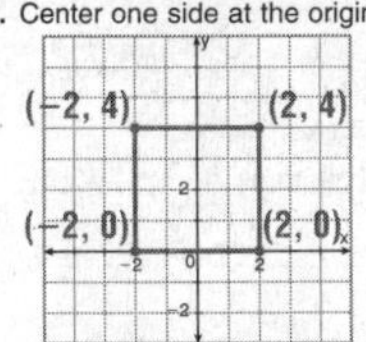

2. Use the origin as a corner. **Possible answer:**

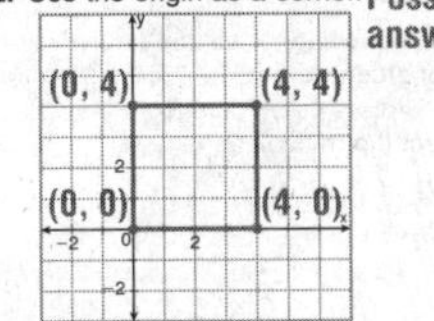

Complete Exercises 3–6 to finish the proof.

Given: Rectangle ABCD has vertices A(0, 4), B(6, 4), C(6, 0), and D(0, 0). E is the midpoint of $\overline{DC}$. F is the midpoint of $\overline{DA}$.

Prove: The perimeter of rectangle DEGF is one half the perimeter of rectangle ABCD.

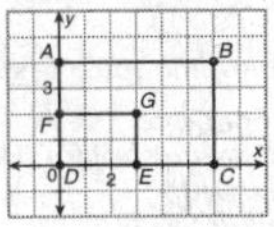

3. Find the perimeter of rectangle ABCD. **20 units**

4. Use the Midpoint Formula to find the coordinates of E and F. Then find the coordinates of G.

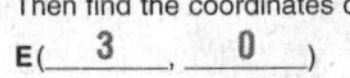  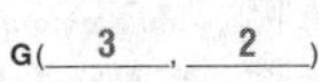

E(**3** , **0**) F(**0** , **2**) G(**3** , **2**)

5. Find the perimeter of rectangle DEGF. **10 units**

6. Is the perimeter of rectangle DEGF one-half the perimeter of rectangle ABCD? **yes**

Complete Exercises 7–11 to finish the same proof but for all rectangles.

Given: ABCD is a rectangle. E is the midpoint of $\overline{DC}$. F is the midpoint of $\overline{DA}$.

Prove: The perimeter of rectangle DEGF is one-half the perimeter of rectangle ABCD.

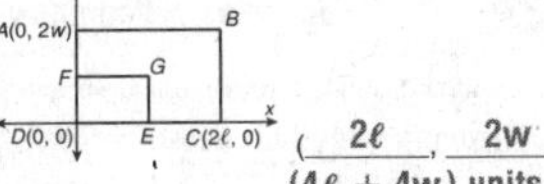

7. Find the coordinates of B. (**2ℓ** , **2w**)

8. Find the perimeter of rectangle ABCD. **(4ℓ + 4w) units**

9. Use the Midpoint Formula to find the coordinates of E and F. Then find the coordinates of G.

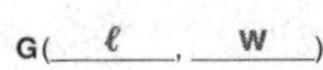

E(**ℓ** , **0**) F(**0** , **w**) G(**ℓ** , **w**)

10. Find the perimeter of rectangle DEGF. **(2ℓ + 2w) units**

11. Is the perimeter of rectangle DEGF one-half the perimeter of ABCD? **yes**

LESSON 4-7

Practice B

Introduction to Coordinate Proof

Position an isosceles triangle with sides of 8 units, 5 units, and 5 units in the coordinate plane. Label the coordinates of each vertex. (Hint: Use the Pythagorean Theorem.)

1. Center the long side on the x-axis at the origin.

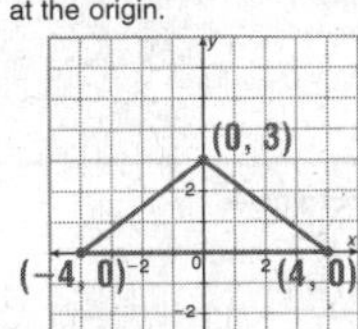

2. Place the long side on the y-axis centered at the origin.

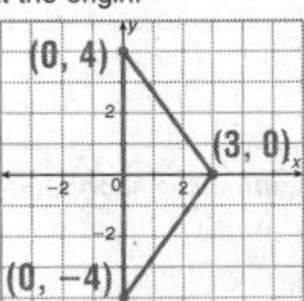

Write a coordinate proof.

3. **Given:** Rectangle *ABCD* has vertices *A*(0, 4), *B*(6, 4), *C*(6, 0), and *D*(0, 0). *E* is the midpoint of $\overline{DC}$. *F* is the midpoint of $\overline{DA}$.

Prove: The area of rectangle *DEGF* is one-fourth the area of rectangle *ABCD*.

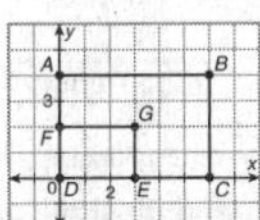

Possible answer: ABCD is a rectangle with width AD and length DC. The area of ABCD is (AD)(DC) or (4)(6) = 24 square units. By the Midpoint Formula, the coordinates of E are $\left(\frac{0+6}{2}, \frac{0+0}{2}\right) = (3, 0)$ and the coordinates of F are $\left(\frac{0+0}{2}, \frac{0+4}{2}\right) = (0, 2)$. The x-coordinate of E is the length of rectangle DEGF, and the y-coordinate of F is the width. So the area of DEGF is (3)(2) = 6 square units. Since $6 = \frac{1}{4}(24)$, the area of rectangle DEGF is one-fourth the area of rectangle ABCD.

LESSON 4-7

Practice C

Introduction to Coordinate Proof

1. A "Yield" sign is an equilateral triangle. Draw an equilateral triangle with side length ℓ in a coordinate plane so that one side is centered at the origin and falls along the x-axis. Determine the coordinates of each vertex in terms of ℓ.

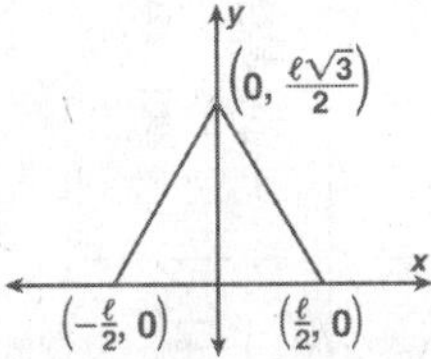

2. A "Stop" sign is a regular octagon. A regular octagon has eight congruent sides and eight congruent 135° angles. The figure shows an octagon with side length ℓ in a coordinate plane so that one side falls along the x-axis and one side falls along the y-axis. Determine the coordinates of each vertex in terms of ℓ. (Hint: You will have to discover a relationship between the sides of the small right triangle at the origin.)

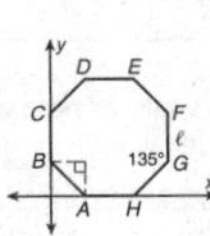

$A\left(\frac{\sqrt{2}}{2}\ell, 0\right)$, $B\left(0, \frac{\sqrt{2}}{2}\ell\right)$, $C\left(0, \frac{\sqrt{2}+2}{2}\ell\right)$, $D\left(\frac{\sqrt{2}}{2}\ell, (\sqrt{2}+1)\ell\right)$, $E\left(\frac{\sqrt{2}+2}{2}\ell, (\sqrt{2}+1)\ell\right)$, $F\left((\sqrt{2}+1)\ell, \frac{\sqrt{2}+2}{2}\ell\right)$, $G\left((\sqrt{2}+1)\ell, \frac{\sqrt{2}}{2}\ell\right)$, $H\left(\frac{\sqrt{2}+2}{2}\ell, 0\right)$

3. A "Slow down" sign is a regular hexagon. A regular hexagon has six congruent sides and six congruent 120° angles. The figure shows a hexagon with side length ℓ in a coordinate plane so that one side falls along the x-axis and one vertex falls on the y-axis. Determine the coordinates of each vertex in terms of ℓ. (Hint: Look back to Exercise 1.)

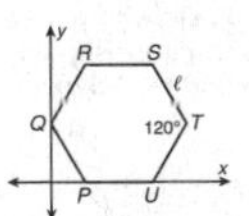

$P\left(\frac{1}{2}\ell, 0\right)$, $Q\left(0, \frac{\sqrt{3}}{2}\ell\right)$, $R\left(\frac{1}{2}\ell, \sqrt{3}\ell\right)$, $S\left(\frac{3}{2}\ell, \sqrt{3}\ell\right)$, $T\left(2\ell, \frac{\sqrt{3}}{2}\ell\right)$, $U\left(\frac{3}{2}\ell, 0\right)$

LESSON 4-7

Reteach

Introduction to Coordinate Proof

A **coordinate proof** is a proof that uses coordinate geometry and algebra. In a coordinate proof, the first step is to position a figure in a plane. There are several ways you can do this to make your proof easier.

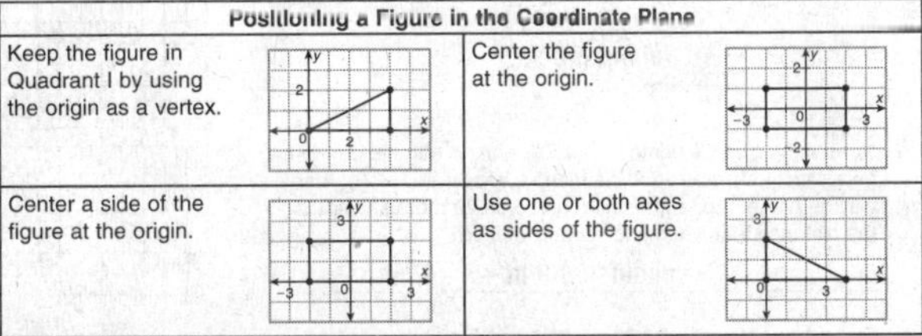

Positioning a Figure in the Coordinate Plane	
Keep the figure in Quadrant I by using the origin as a vertex.	Center the figure at the origin.
Center a side of the figure at the origin.	Use one or both axes as sides of the figure.

Position each figure in the coordinate plane and give the coordinates of each vertex.

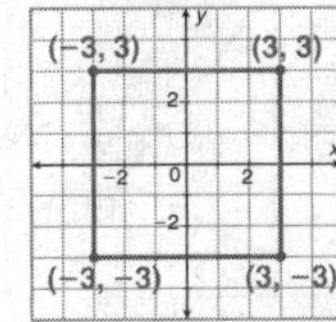

1. a square with side lengths of 6 units

Possible answer on graph above.

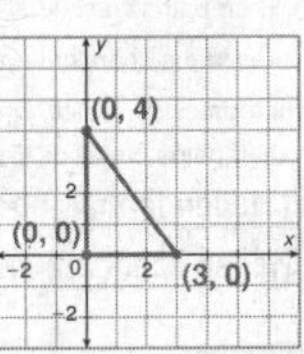

2. a right triangle with leg lengths of 3 units and 4 units

Possible answer on graph above.

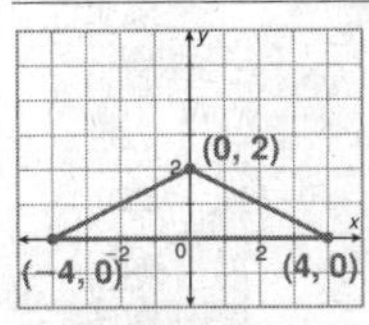

3. a triangle with a base of 8 units and a height of 2 units

Possible answer on graph above.

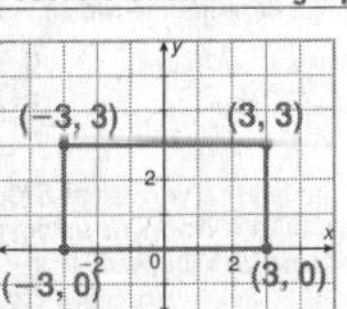

4. a rectangle with a length of 6 units and a width of 3 units

Possible answer on graph above.

LESSON 4-7

Reteach

Introduction to Coordinate Proof *continued*

You can prove that a statement about a figure is true without knowing the side lengths. To do this, assign variables as the coordinates of the vertices.

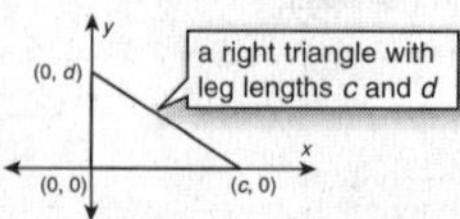

Position each figure in the coordinate plane and give the coordinates of each vertex.

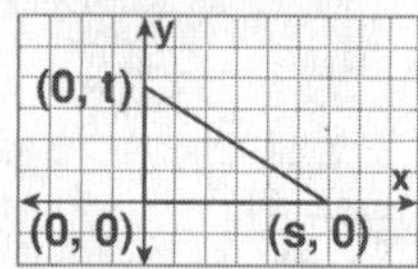

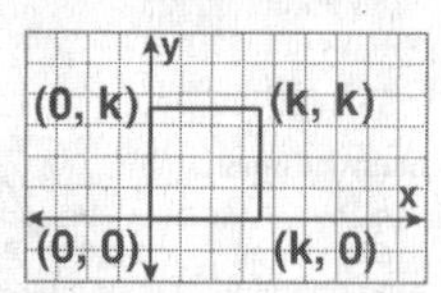

5. a right triangle with leg lengths *s* and *t*

   Possible answer on graph above.

6. a square with side lengths *k*

   Possible answer on graph above.

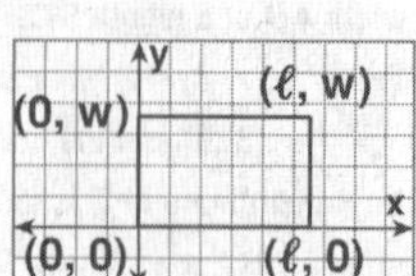

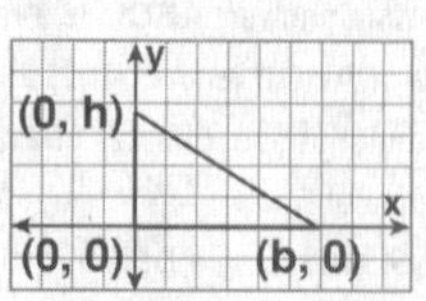

7. a rectangle with leg lengths ℓ and *w*

   Possible answer on graph above.

8. a triangle with base *b* and height *h*

   Possible answer on graph above.

9. Describe how you could use the formulas for midpoint and slope to prove the following.

   Given: $\triangle HJK$, *R* is the midpoint of $\overline{HJ}$, *S* is the midpoint of $\overline{JK}$.

   Prove: $\overline{RS} \parallel \overline{HK}$

   Possible answer: Use the midpoint formula to find the coordinates of the midpoints R and S. Then use the coordinates and the formula for slope to find the slopes of $\overline{RS}$ and $\overline{HK}$.

LESSON 4-7

Challenge

Using Coordinate Proof to Verify Conjectures

A **parallelogram** is a quadrilateral with both pairs of opposite sides parallel. If one pair of opposite sides of a quadrilateral are both parallel and congruent, then the quadrilateral is a parallelogram.

1. Draw a quadrilateral so that no two sides are congruent or parallel. Use a ruler to find and draw the midpoint of each side. Connect the midpoints.

   Possible drawing:

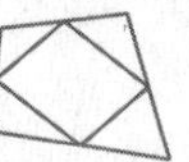

2. Make a conjecture about the figure that is formed when the consecutive midpoints are connected.

   Possible answer: A parallelogram is formed.

Use the figure and the information below to answer each question.

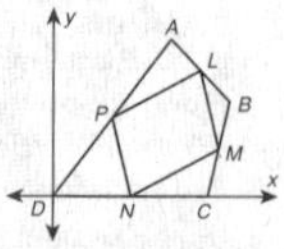

Given: *ABCD* is a quadrilateral. *L, M, N,* and *P* are midpoints of sides $\overline{AB}$, $\overline{BC}$, $\overline{CD}$, and $\overline{DA}$, respectively.

Prove: *LMNP* is a parallelogram.

3. Give the coordinates of each vertex of quadrilateral *ABCD* as it is positioned in the coordinate plane. (*Hint:* Use coordinates that are multiples of 2 to make finding the midpoints easier.)

   Possible answer: A(2a, 2d), B(2b, 2e), C(2c, 0), D(0, 0)

4. Use the midpoint formula to find the coordinates of *L, M, N,* and *P.*

   Answers based on coordinates in Exercise 3: L(a + b, d + e), M(b + c, e), N(c, 0), P(a, d)

5. Find the slopes of $\overline{PL}$ and $\overline{MN}$.

   Answers based on coordinates in Exercise 4: $\frac{e}{b}$, $\frac{e}{b}$

6. Use the distance formula to find *PL* and *MN.*

   Answers based on coordinates in Exercise 4: $\sqrt{b^2 + e^2}$; $\sqrt{b^2 + e^2}$

7. What can you conclude?

   $\overline{PL}$ and $\overline{MN}$ have the same slope, so they are ∥. $\overline{PL}$ and $\overline{MN}$ also have the same length, so they are ≅. Quadrilateral LMNP is a parallelogram because if one pair of opposite sides of a quadrilateral is both ∥ and ≅, then the quadrilateral is a parallelogram.

LESSON 4-7

Problem Solving

Introduction to Coordinate Proof

Round to the nearest tenth for Exercises 1 and 2.

1. A fountain is at the center of a square courtyard. If one grid unit represents one yard, what is the distance from the fountain at (0, 0) to each corner of the courtyard?

   about 4.2 yd

2. Noah started at his home at *A*(0, 0), walked with his dog to the park at *B*(4, 2), walked to his friend's house at *C*(8, 0), then walked home. If one grid unit represents 20 meters, what is the distance that Noah and his dog walked?

   about 338.9 m

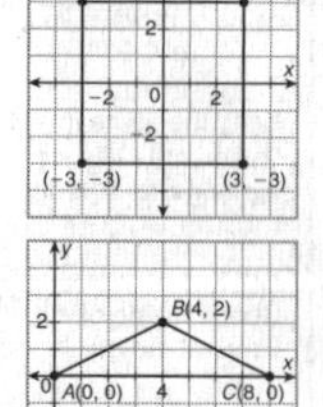

Use the following information for Exercises 3 and 4.

Rachel started her cycling trip at *G*(0, 7). Malik started his trip at *J*(0, 0). Their paths crossed at *H*(4, 2).

3. Draw their routes in the coordinate plane.

4. If one grid unit represents $\frac{1}{2}$ mile, who had ridden farther when their paths crossed? Explain.

   Rachel had ridden farther, about 3.2 miles; Malik had ridden about 2.2 miles.

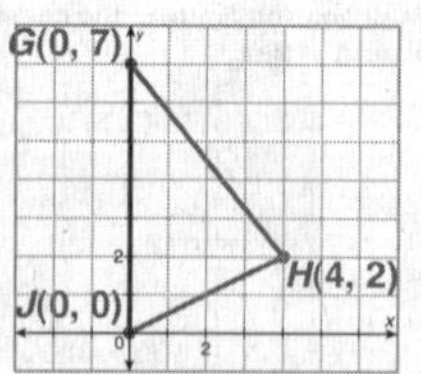

Choose the best answer.

5. Two airplanes depart from an airport at *A*(9, 11). The first airplane travels to a location at *N*(–250, 80), and the second airplane travels to a location at *P*(105, –400). Each unit represents 1 mile. What is the distance, to the nearest mile, between the two airplanes?

   A 335.3 mi
   C 490.3 mi
   B 477.9 mi
   (D) 597.0 mi

6. A corner garden has vertices at *Q*(0, 0), *R*(0, 2*d*), and *S*(2*c*, 0). A brick walkway runs from point *Q* to the midpoint *M* of $\overline{RS}$. What is *QM*?

   F (*c, d*)
   H $\sqrt{c + d}$
   G $c^2 + d^2$
   (J) $\sqrt{c^2 + d^2}$

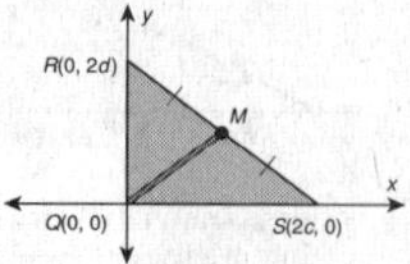

LESSON 4-7

Reading Strategies

Synthesize Information

Figures can be positioned in a coordinate plane in one of four ways:

Use the **origin as a vertex,** which may keep the figure in Quadrant I.	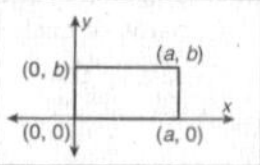 (0, b), (a, b), (0, 0), (a, 0)
Center the figure at the origin of the coordinate plane.	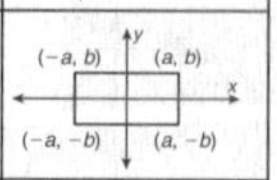 (−a, b), (a, b), (−a, −b), (a, −b)
Center a side of the figure at the origin of the coordinate plane.	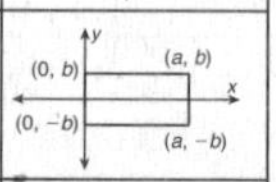 (0, b), (a, b), (0, −b), (a, −b)
Use **one or both axes** as sides of the figure.	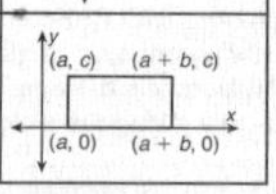 (a, c), (a + b, c), (a, 0), (a + b, 0)

Using the given information, position the figure on the coordinate plane provided and answer the following questions.

1. Where would you position a triangle on a coordinate plane if you want to find the area of the triangle?

   with one side along one of the axes

2. Where would you position a triangle to find the midpoint of a side?

   with the side you want to find the midpoint of along an axis with the midpoint at the origin

Indicate where on a coordinate plane each figure should be placed in order to find the following measurement.

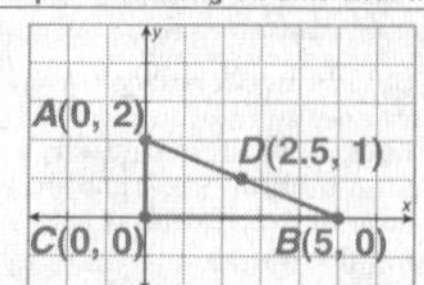

3. the area of $\triangle ABC$ — **with one side along an axis**
4. the midpoint of $\overline{AB}$ — **with $\overline{AB}$ straddling the origin on an axis**
5. the area of $\triangle ADC$ — **with one vertex at the origin and one side on an axis**
6. the area of $\triangle CDB$ — **with one vertex at the origin and one side on an axis**

LESSON 4-8

Practice A

Isosceles and Equilateral Triangles

Name the parts of the figure that match the vocabulary words.

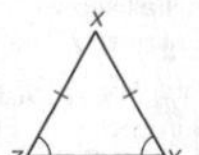

1. base: $\overline{ZY}$
2. legs: $\overline{XY}$ and $\overline{XZ}$
3. base angles: $\angle Z$ and $\angle Y$
4. vertex angle: $\angle X$

Fill in the blanks in Exercises 5–8 to complete each theorem.

5. If a triangle is equilateral, then it is **equiangular**.
6. If two angles of a triangle are congruent, then the sides **opposite** those angles are congruent.
7. If two sides of a triangle are congruent, then the **angles** opposite those sides are congruent.
8. If a triangle is equiangular, then it is **equilateral**.
9. A forest ranger in Grand Canyon National Park wants to find the minimum distance across the canyon. She finds a place in the Marble Canyon area of the park where the sides seem close together. She takes measurements and draws this figure. Find the distance AB. (Hint: The angles in an equiangular triangle measure 60°.) **693 ft**

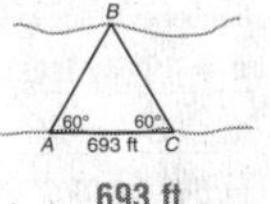

Find each value.

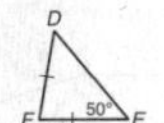

10. m∠D = **50°**

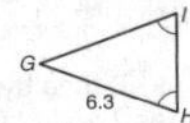

11. GI = **6.3**

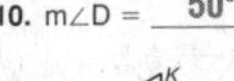

12. m∠L = **60°**

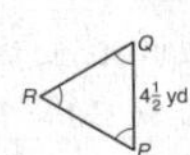

13. RQ = **$4\frac{1}{2}$ yd**

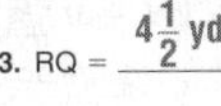

14. m∠U = **65°**

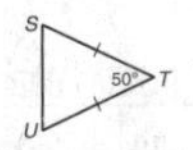

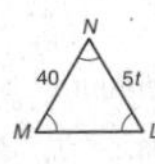

15. t = **8**

LESSON 4-8

Practice B

Isosceles and Equilateral Triangles

An altitude of a triangle is a perpendicular segment from a vertex to the line containing the opposite side. Write a paragraph proof that the altitude to the base of an isosceles triangle bisects the base.

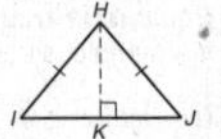

1. **Given:** $\overline{HI} \cong \overline{HJ}$, $\overline{HK} \perp \overline{IJ}$

   Prove: $\overline{HK}$ bisects $\overline{IJ}$.

   Possible answer: It is given that $\overline{HI}$ is congruent to $\overline{HJ}$, so ∠I must be congruent to ∠J by the Isosceles Triangle Theorem. ∠IKH and ∠JKH are both right angles by the definition of perpendicular lines, and all right angles are congruent. Thus by AAS, △HKI is congruent to △HKJ. $\overline{IK}$ is congruent to $\overline{KJ}$ by CPCTC, so $\overline{HK}$ bisects $\overline{IJ}$ by the definition of segment bisector.

2. An obelisk is a tall, thin, four-sided monument that tapers to a pyramidal top. The most well-known obelisk to Americans is the Washington Monument on the National Mall in Washington, D.C. Each face of the pyramidal top of the Washington Monument is an isosceles triangle. The height of each triangle is 55.5 feet, and the base of each triangle measures 34.4 feet. Find the length, to the nearest tenth of a foot, of one of the two equal legs of the triangle. **58.1 ft**

Find each value.

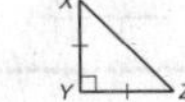

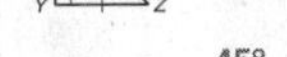

3. m∠X = **45°**

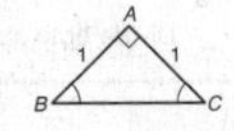

4. BC = **$\sqrt{2}$**

5. PQ = **36 or 9**

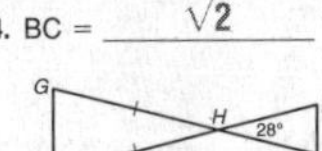

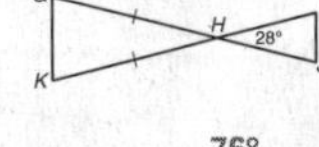

6. m∠K = **76°**

7. t = **$\frac{4}{3}$**

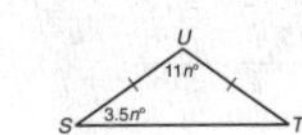

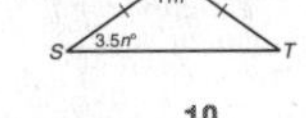

8. n = **10**

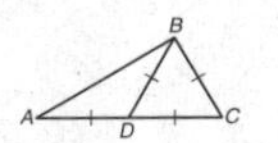

9. m∠A = **30°**

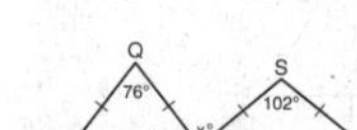

10. x = **89**

LESSON 4-8

Practice C

Isosceles and Equilateral Triangles

Draw a figure, name the coordinates, and write a coordinate proof.

1. **Given:** △ABC is an isosceles triangle. D is the midpoint of the base $\overline{BC}$.

   Prove: $\overline{AD} \perp \overline{BC}$

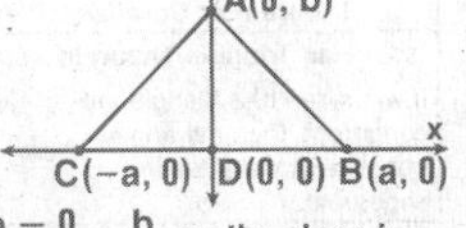

Possible answer: △ABC is an isosceles triangle with vertices A(0, b), B(a, 0), and C(−a, 0). D is the midpoint of $\overline{BC}$, so D has coordinates (0,0). The slope of $\overline{AD}$ is $\frac{b-0}{0-0} = \frac{b}{0}$, so the slope is undefined. A line with an undefined slope is a vertical line. The slope of $\overline{BC}$ is $\frac{0-0}{a-(-a)} = \frac{0}{2a} = 0$. A line with a zero slope is a horizontal line. Because $\overline{AD}$ is vertical and $\overline{BC}$ is horizontal, $\overline{AD} \perp \overline{BC}$.

A soccer ball is covered by a pattern of regular pentagons and regular hexagons. On a flat surface (like this page), the edges of the shapes do not meet exactly, but on a sphere (like a ball), they cover the entire shape without any gaps. Use the figure for Exercises 2 and 3.

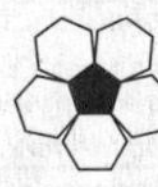

2. A regular pentagon has five congruent sides and five 108° angles. Use the figure and your knowledge of isosceles triangles to find the values of x, y, and z.

   x = 36; y = 72; z = 36

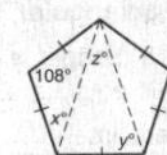

3. A regular hexagon has six congruent sides and six 120° angles. If a regular hexagon has side lengths of ℓ, use the figure and your knowledge of isosceles triangles to find the length of diagonal $\overline{FC}$.

   2ℓ

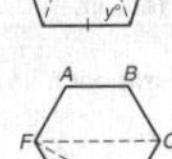

Name the coordinates of vertex Y. Write a coordinate proof.

4. **Given:** △XYZ is an equilateral triangle. A is the midpoint of $\overline{XY}$. B is the midpoint of $\overline{YZ}$. C is the midpoint of $\overline{XZ}$.

   Prove: △ABC, △XAC, △YAB, and △CBZ are congruent equilateral triangles.

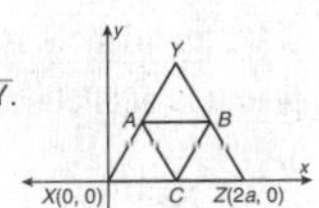

Possible answer: The coordinates of Y are $(a, \sqrt{3}a)$. The Midpoint Formula shows that the midpoints of the sides of △XYZ are $A\left(\frac{1}{2}a, \frac{\sqrt{3}}{2}a\right)$, $B\left(\frac{3}{2}a, \frac{\sqrt{3}}{2}a\right)$, and C(a, 0). The Distance Formula gives these distances: AX = AY = AC = AB = BC = XC = BY = BZ = CZ = a. Thus by SSS, △ABC ≅ △XAC ≅ △YAB ≅ △CBZ, and the triangles are equilateral.

LESSON 4-8

Reteach

Isosceles and Equilateral Triangles

Theorem	Examples
Isosceles Triangle Theorem If two sides of a triangle are congruent, then the angles opposite the sides are congruent.	R, T, S — If $\overline{RT} \cong \overline{RS}$, then ∠T ≅ ∠S.
Converse of Isosceles Triangle Theorem If two angles of a triangle are congruent, then the sides opposite those angles are congruent.	L, N, M — If ∠N ≅ ∠M, then $\overline{LN} \cong \overline{LM}$.

You can use these theorems to find angle measures in isosceles triangles.

Find m∠E in △DEF.

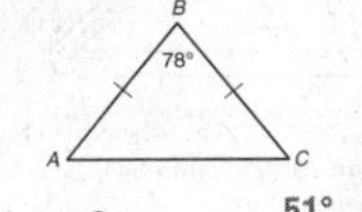

m∠D = m∠E	Isosc. △ Thm.
5x° = (3x + 14)°	Substitute the given values.
2x = 14	Subtract 3x from both sides.
x = 7	Divide both sides by 2.

Thus m∠E = 3(7) + 14 = 35°.

Find each angle measure.

1. m∠C = **51°**

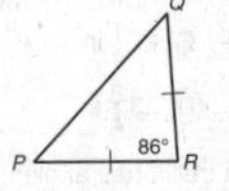

2. m∠Q = **47°**

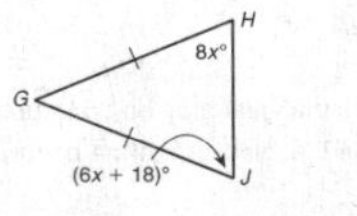

3. m∠H = **72°**

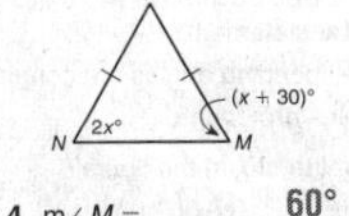

4. m∠M = **60°**

LESSON 4-8

Reteach

Isosceles and Equilateral Triangles continued

Equilateral Triangle Corollary
If a triangle is equilateral, then it is equiangular.

(equilateral △ → equiangular △)

Equiangular Triangle Corollary
If a triangle is equiangular, then it is equilateral.

(equiangular △ → equilateral △)

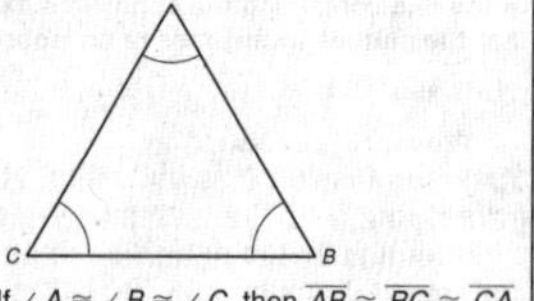

If ∠A ≅ ∠B ≅ ∠C, then $\overline{AB} \cong \overline{BC} \cong \overline{CA}$.

You can use these theorems to find values in equilateral triangles.

Find x in △STV.

△*STV* is equiangular.	Equilateral △ → equiangular △
$(7x + 4)° = 60°$	The measure of each ∠ of an equiangular △ is 60°.
$7x = 56$	Subtract 4 from both sides.
$x = 8$	Divide both sides by 7.

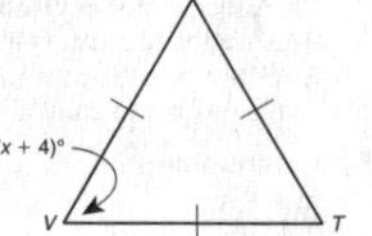

Find each value.

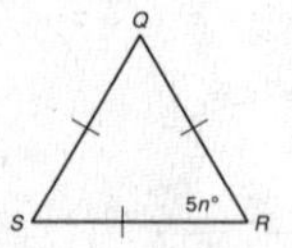

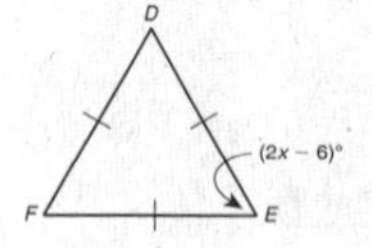

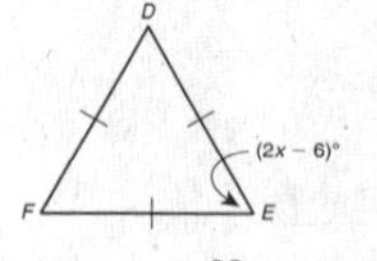

5. *n* = **12**

6. *x* = **33**

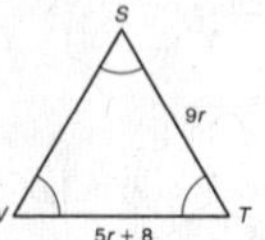

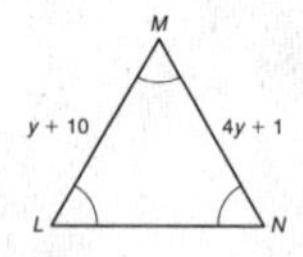

7. *VT* = **18**

8. *MN* = **13**

LESSON 4-8

Challenge

Triangles in Geodesic Domes

A **geodesic dome** can use equilateral triangles to create a super-strong almost spherical structure. Of all known structures, the geodesic dome encloses the greatest amount of space with the least amount of materials.

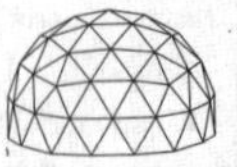

Geodesic domes have various *frequencies*, which are related to the number of smaller triangles in each unit. Four units with different frequencies are shown below.

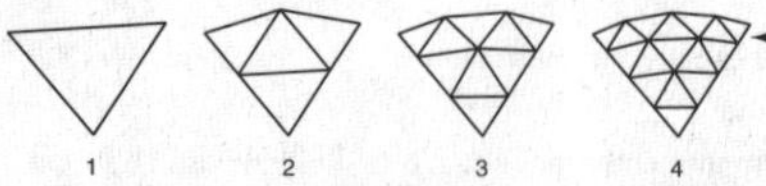

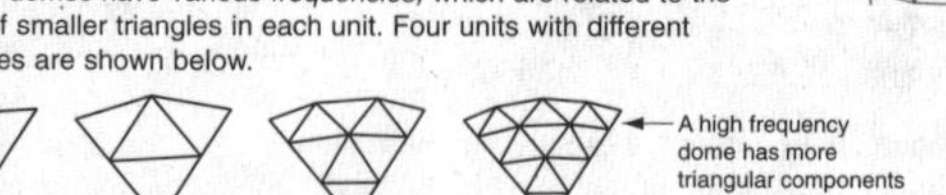

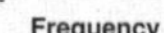

In perspective drawings of geodesic domes, some of the equilateral triangles are shown as isosceles triangles.

1. Given line segments $\overline{PT}$ and $\overline{TR}$, use the diagram of the frequency 2 unit to find *x*. Explain your reasoning.

   x = 10; By Isosc. △ Thm. and △ Sum Thm., m∠TSV = 54°. △QSV ≅ △TSV by SSS, so m∠VSQ = 54° by CPCTC. m∠QSR = 72° by Lin. Pair Thm. and by ∠ Add. Post. By Isosc. △ Thm. and △ Sum Thm., m∠SQR = 36°. Solve (3x + 6)° = 36°. x = 10

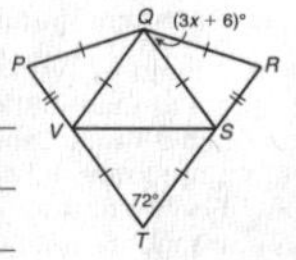

Given line segments $\overline{AK}$ and $\overline{KD}$, use the diagram of the frequency 3 unit for Exercises 2 and 3. Describe the strategies that you used.

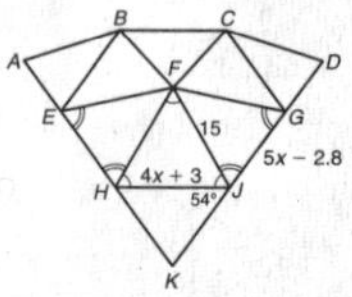

2. Find the perimeter of △*EFH*.

   42.2; △FJH is an equilat. △, so HJ = JF = FH = 15 and x = 3. So GJ = 12.2. $\overline{FG} \cong \overline{FJ}$ since the sides opp. the ∠s are ≅. So FG = FJ = 15. △EFH ≅ △GFJ by AAS, so the corr. sides are ≅ by CPCTC. P = 12.2 + 15 + 15 = 42.2

3. Find m∠*GFH*.

   108°; △FJH is equiangular, so m∠JFH = m∠FJH = 60°. By Lin. Pair Thm. and ∠ Add. Post., m∠FJG = 66°. By Isosc. △ Thm. and △ Sum Thm., m∠GFJ = 48°. m∠GFH = 48° + 60° = 108°

LESSON 4-8

Problem Solving

Isosceles and Equilateral Triangles

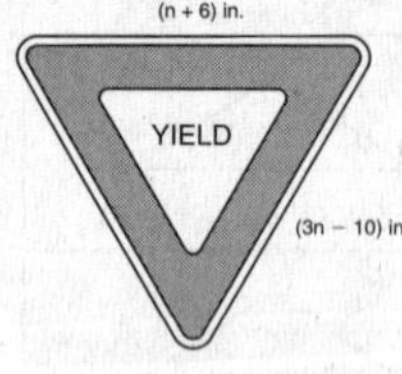

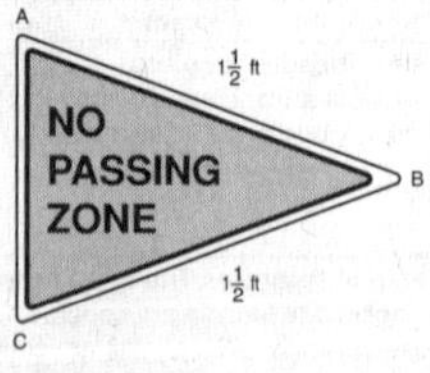

1. A "Yield" sign is an equiangular triangle. What are the lengths of the sides?

   14 in.

2. The measure of ∠C is 70°. What is the measure of ∠B?

   40°

3. Samantha is swimming along $\overrightarrow{HF}$. When she is at point H, she sees a necklace straight ahead of her but on the bottom of the pool at point J. Then she swims 11 more feet to point G. Use the diagram to find GJ, the distance Samantha is from the necklace. Explain.

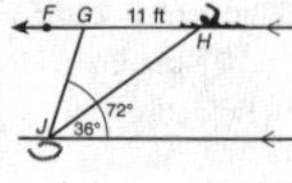

   11 ft; m∠GJH = 72° − 36° = 36°. m∠GHJ = 36° by Alt. Int. ∠s Thm. By Converse of Isosceles Triangle Theorem, GJ = GH = 11 ft.

Choose the best answer.

4. A billiards triangle is equiangular. What is the perimeter?

   A $5\frac{1}{8}$ in.
   C $11\frac{1}{4}$ in.
   B $10\frac{1}{4}$ in.
   (D) $33\frac{3}{4}$ in.

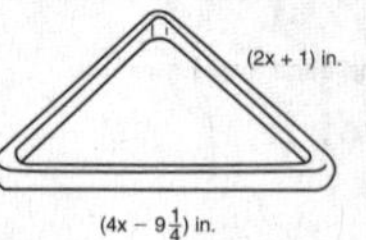

5. A triangular shaped trellis has angles R, S, and T that measure 73°, 73°, and 34°, respectively. If ST = 4y + 6 and TR = 7y − 21, what is the value of y?

   F 5
   H 11
   (G) 9
   J 15

6. Two triangular tiles each have two sides measuring 4 inches. Which is a true statement?

   A Their corresponding angles are congruent.
   (C) The triangles may be congruent.
   B The triangles are congruent.
   D The triangles cannot be congruent.

7. What is the value of x in the figure?

   F 42°
   (H) 96°
   G 90°
   J 106°

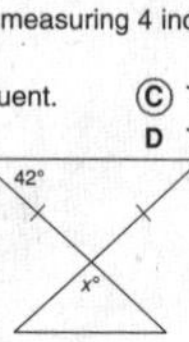

LESSON 4-8

Reading Strategies

Understanding Relationships

Isosceles and equilateral triangles can be described in the following ways.

Theorem or Corollary	Hypothesis and Conclusion	Example
Isosceles Triangle Theorem If two sides of a triangle are congruent, then the angles opposite those sides are congruent.	If $\overline{XZ} \cong \overline{XY}$, then ∠Y ≅ ∠Z.	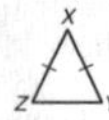
Converse of Isosceles Triangle Theorem If two angles of a triangle are congruent, then the sides opposite those angles are congruent.	If ∠N ≅ ∠M, then $\overline{LM} \cong \overline{LN}$.	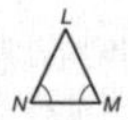
Equilateral Triangle Corollary If a triangle is equilateral, then it is equiangular. (equilateral △ → equiangular △)	If $\overline{QR} \cong \overline{RS} \cong \overline{SQ}$, then ∠Q ≅ ∠R ≅ ∠S.	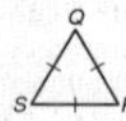
Equiangular Triangle Corollary If a triangle is equiangular, then it is equilateral. (equilateral △ → equiangular △)	If ∠E ≅ ∠F ≅ ∠G, then $\overline{EF} \cong \overline{FG} \cong \overline{GE}$.	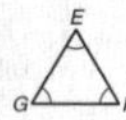

Find each value and indicate which theorem you used in determining the answer.

1. 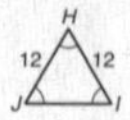 *JI* **12; If a triangle is equiangular, then it is equilateral.**

2. 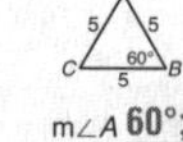

   m∠*A* **60°; If a triangle is equilateral, then it is equiangular.**

3. 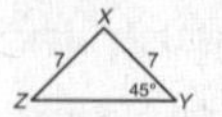 m∠*Z* **45°; If two sides of a triangle are congruent, then the angles opposite those sides are congruent.**

4. 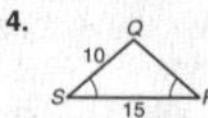

   QR **10; If two angles of a triangle are congruent, then the sides opposite those angles are congruent.**